Enclosing Water

Stefania Barca

Enclosing Water

Nature and Political Economy in a Mediterranean Valley,
1796–1916

The White Horse Press

British Library Cataloguing in Publication Data
A catalogue record for this book is available from the British Library

ISBN   978-1-874267-56-0 (HB); 978-1-874267-57-7 (PB)

To Marco and Giulia

Contents

Abbreviations

Some archival sources are referred to in the notes using the following abbreviations. Translations from these and other Italian language sources are the author's own, unless otherwise stated.

AC (Affari Comunali)
ACSR (Archivio Centrale dello Stato-Roma)
ADC (Atti Diversi Comunali)
AIC (Agricoltura Industria e Commercio)
AP (Alta Polizia)
ASC (Archivio di Stato di Caserta),
ASN (Archivio di Stato di Napoli),
BSCN (*Bollettino delle Sentenze della Commissione Feudale*)
CA (Contenzioso Amministrativo)
CPTL (Consigli Provinciali in Terra di Lavoro)
DGPAS (Direzione Generale Ponti, Acque e Strade)
IAC (Intendenza Affari Comunali)
IB (Intendenza Borbonica)
Inv. (inventario)
LLPP (Ministero dei Lavori Pubblici)
MI (Ministero dell'Interno)
PC (Prefettura Contratti)
PCA (Prefettura, Carte Amministrative)
PPS (Prefettura Prima Serie)
PS (Ponti e Strade)
PS-B (Ponti e Strade-Bonifiche)

Biographical Note

Stefania Barca (b. Naples 1968) is a researcher at the Centre for Social Studies of the University of Coimbra in Portugal. After obtaining her PhD in Economic History in 1997, she collaborated with the Institute of Studies on Mediterranean Societies in Naples and lectured in economic, social and environmental history at various Universities in Italy. She was a visiting scholar on the Program in Agrarian Studies at Yale University in 2005–06, and a Ciriacy Wantrup post-doctoral fellow at the University of California, Berkeley in 2006–08. Her publications include the books *Elettrificare la Puglia* (2001, 'The Electrification of Apulia') and *Storia dell'ambiente. Una introduzione* (2004, 'An Introduction to Environmental History'), with Marco Armiero. Her next research projects are a transnational environmental history of petrochemicals, and a biography of Italian ecologist Laura Conti.

Acknowledgments

This book was born at the end of a long journey, both intellectual and geographical. I wish to thank the many people and institutions that have helped me on my way.

Early support for the research in Naples was granted by the Institute of Studies on Mediterranean Societies, University of Naples 'l'Orientale' and University of Naples Federico II. For those grants, I am very grateful to Gabriella Corona, Paolo Frascani, Vincenzo Giura, Paolo Malanima and Ilaria Zilli. That early research became an English manuscript in the course of three years which I spent in the US, first as visiting scholar at the Yale University Program in Agrarian Studies and then as S.V. Ciriacy Wantrup postdoctoral fellow at UC Berkeley. I am especially indebted to Jim Scott, for introducing my mind to an entire new world of intellectual possibilities; and to Carolyn Merchant, for her radical way of thinking ecology and for her maternal friendship. Many other people welcomed, sustained and helped me to get the best from my stay in the US, and to all I am truly thankful. I wish especially to thank Kay Mansfield for her precious support during my first year in Connecticut; Christine Meisner Rosen, Nancy Peluso, Isha Ray and Michael Watts for making possible and indeed very pleasant my stay in Berkeley and for the intellectual support I enjoyed from all them. Myrna Santiago, Angus Wright and Mary Mackey also warmly welcomed me and my family in California, and I am very happy I had the opportunity of making friends with them.

I am also deeply indebted to Piero Bevilacqua, for the very idea of the topic of this book and for continued support down the years; and to Donald Worster, for the great privilege of his friendship and for the inspiration which I have enjoyed from him over many years. I am very grateful to Joan Martinez Alier and Marta Petrusewicz for the precious advice and support they granted me on many occasions; to John McNeill and to Nancy Peluso for having read provisory, often immature, pieces of this work and for the help they generously gave me in thinking it through. Part of Chapter 3 was first presented at the Berkeley Environmental Politics Workshop in November 2007 and is now included in *Nature and History in Modern Italy*, edited by Marco Armiero and Marcus Hall (Ohio University Press): I wish to thank Bern Gardener for his insightful commentary and all the participants for their appreciations and suggestions. Thanks also to the Land-Lab people of UC Berkeley in 2007–08 for their comments on other parts of the chapter. Gabrielle Bouleau, Costanza D'Elia, Rohan D'Souza, Marcus Hall, Giorgos Kallis, Paolo Macry, Marta Petrusewicz, have also read and commented upon pieces of this work and I have enjoyed their insights and advice. Special thanks go to Rosaria Bellucci, Gianluca Cellupica, Ferdinando Corradini, Silvio De Majo, Giulio De Martino and Stefano Mancini for facilitating my research in Isola del Liri.

Acknowledgments

The Centre for Social Studies of the University of Coimbra generously granted additional funds for the English revision of the text: special thanks go to Mark Walters for his conscientious job and competent suggestions about river terminology.

Last, but not least, I wish to thank my parents, Renata and Lorenzo, for their long-term, long-distance and unconditional support.

This book is dedicated to Marco Armiero, who has shared with me this long journey through the Liri Valley and through life; and to Giulia, for the great adventure I have enjoyed meanwhile: that of growing with her.

Enclosing Water

Introduction

We need different ideas because we need different relationships
Raymond Williams, 'Ideas of Nature'

I

This book tells the story of the Industrial Revolution – probably the greatest transformation of society–nature relationships in the modern era – from the vantage point of a place called the Liri Valley: a roughly 10-mile region of gently sloping hills, watered by the River Liri and its tributary Fibreno, nestled within central Italy's Apennines between Naples and Rome. The book looks at industrialisation from the banks of the river that moved the wheels of a factory system and from the standpoint of the people inhabiting and working the place[1].

Though challenged by many scholars and undermined by post-modernist critiques of universal meta-narratives, the concept of Industrial Revolution is maintained here and reinterpreted in socio-ecological fashion. Of course, the idea of the Industrial Revolution as a unique and progressive paradigm of human history is faulty and does not account for the diversity of places, cultures and historical conditions[2]. Nevertheless, the idea that history is fragmented into a multitude of localities and differences also seems inadequate to make sense of the changes occurring in these same places. Political economy, the 'large scale configuration of power', as Donald Worster argues, is needed if we want to make sense of the connections between different scales and dimensions of society-nature interaction through time[3]. This book offers insights into how the global political economy of industrialisation and the natural environment of a particular river valley in the Mediterranean constituted each other, producing a new landscape and social formation: the 'Manchester of the Two Sicilies' – as the Liri Valley was called in the mid-1840s[4].

The book will present a prospect on a marginal Apennine area on Europe's periphery and show how it was connected to the ideal and material reality of nineteenth century industrialisation. Rather than an account of how the natural environment – an object difficult of definition in itself – was altered and at times destroyed by humans, this is a reflection on how human history is *also* the history of the environment, for it embodies in particular places and is made of them. The place, with its ecology and culture, will play a huge part in the unfolding of the story and will give us an unfamiliar perspective on the Industrial Revolution: that of a Mediterranean watershed.

II

When I started this investigation, the industrial transformation of the Liri Valley appeared to me the result of a simple equation: waterpower plus private property – an age-old technology on which new social relationships had been imposed. This historical process was part of what my primary sources (government files in the Bourbon Archives) called the 'Economy of Water' – meaning the many ways in which water could be added to labour to increase the value of production[5]. Soon enough, however, I realised that the 'Economy of Water' not only meant increased production; transforming water into natural capital also implied less positive consequences for the environment and for the people inhabiting it. The most striking of these unintended consequences was hydrological risk, especially flooding.

Probably the most widely recognised environmental threat in the nineteenth century Mediterranean, floods have been mostly related to land use changes, mainly deforestation. In the 1790s, the philosophers of the Neapolitan Enlightenment school developed a theory of deforestation and environmental risk in the Mediterranean: what they called 'the disorder of water'. They were convinced that it was political change – namely, the Barbarian invasions, bringing about the feudal–communal system of land tenure – that had caused environmental and particularly hydrological instability. And they envisioned political economy as the best means to restore the natural harmony between humans and their environment in the country. Such a mix of ideas and myths about nature, politics, history and the geography of southern Italy formed the cultural substrate for the enclosure of the Liri and the industrial transformation of its valley[6].

Environmental historians of southern Italy have generally retained the concept of 'disorder of water' for its ability to connect land and water processes, mountains and plains, forests and agriculture within a unique socio-environmental explanation[7]. Hydrological instability in the nineteenth century Apennines has been generally interpreted as the result of mixed natural and social causes: the effect of the torrential regime of watercourses combined with population growth and the advent of agrarian capitalism since 1750. This vision is centred on the fundamental assumption that, in a country so densely and long inhabited as Italy, one cannot talk of a 'nature' to preserve in a pristine state: the environment is the historical product of past human work and social organisation coalesced with the acting of natural forces. Lying within the Mediterranean climate and formed of geologically young mountains, Italy's nature is inherently disorderly when measured against human habitation: earthquakes, landslides and floods still play a major role among the 'original characters' of the land[8]. The point in Italian environmental history seems to be how much social agency has stewarded natural processes and counterbalanced risk, or else increased it with short-sighted choices. Most of all, environmental risk is seen as a political economy issue and floods and malaria are blamed on the failure of the State to articulate private interest with the public good[9].

Introduction

What the debate has mostly overlooked, however, are the changes in water property and use – in particular the enclosure of rivers into the factory system – and the way in which these might have interlaced with changes in land use to increase the risk of flooding.

In a sense, this book forms part of a major current in environmental history, concerned with rivers in the age of industry. Scholars in this field have devoted much effort to analysing the changes in water use which were crucial to industrialisation itself, producing a number of valuable insights into how, by transforming rivers, modern societies have transformed themselves[10]. They have shown how industrial societies have used water to sustain an unprecedented development of human wealth, energy potential, urban life, mobility etc, and described rivers as a major victim of this modern growth. They have shown how water has been reconfigured in the process to the point that most of the Earth's rivers have become unrecognisable with respect to their pre-industrial appearance. Indeed, more than at any time in past centuries, the last two hundred years have witnessed the war of humans on the 'disorder of water' to sustain the socio-economic order of capitalist agriculture, urban life and industrialisation. Moors and lakes have been drained, rivers dammed, channelled and embanked on a scale unprecedented in past societies. In turn, no urban–industrial way of life as we know it today could be imaginable without such widespread redesigning of watersheds[11].

Environmental risk as related to water use, especially in urban contexts, is a central concern of many of the above studies. The uneven production of risk emerges as a common feature of industrialised watersheds: having been re-engineered to suit industry and urban life, and despite huge efforts to master water flow, rivers have inexplicably refused to act in an orderly manner and continue to unleash major floods to this day. Moreover, such unnatural disasters tend to reflect the distribution of power and wealth within society. Thus they hit some people more than others and can even benefit some groups in the end. Floods seem to obey the logic of the social costs of private enterprise[12].

The Liri participated in this process of reconfiguring the Earth's rivers in the age of industry with its own peculiarities, natural and social: those of an Apennine watercourse in a proto-industrial region of the European South. Viewing the transformation of this particular river as a historical subject with its own story, not entirely covered by the deforestation-soil erosion narrative, offers a fresh perspective on environmental risk in the nineteenth century Mediterranean. Albeit a minor phenomenon, industrialisation *did* happen in the Italian Apennines and, until the mid-twentieth century, was almost entirely a waterpower affair. What this transformation of rural into industrial meant for many river valleys along the peninsula and for their people is not unknown to historians; but just how inextricably social change was linked to environmental change across different scales – the place, the nation and the global economy of the time – is still a largely unwritten story.

Introduction

This book argues that, in redefining the river as property, and materially enclosing it in the factory system, industrial capitalists contributed to producing hydrological risk on a scale and with an intensity only comparable to that of mechanised production. Indeed, they ended up manufacturing floods by the same means with which they were manufacturing woollens and paper sheets. This was an early and indeed striking manifestation of what ecological Marxists call the 'second contradiction' of capitalism, that between nature and the logic of capital accumulation. In a marginal rural economy of the Mediterranean Apennines, the possession of waterpower was more valuable than that of land: once the very symbol of feudal power and the greatest source of feudal revenue, the energy of the river was now freely available to capital. Wherever possible, landowners aspired to become industrialists, to possess their own waterpower by claiming exclusive rights over 'enclosed' parts of the river. The 'second contradiction' thus appears in the Liri Valley as one between flow and stock, between the fluid nature of the river and the abstract logic of accumulation superimposed on it. The book looks at this remaking of water into property as the very essence of the Industrial Revolution in the Liri Valley and investigates the unequal relationships of access and vulnerability brought about in the process[13].

III

One last word is needed to introduce the present study: this is a human-centred environmental history. Among many species inhabiting the place in which the story unfolds, the book focuses on humans and is built around their world: a world in which the struggle for the control of nature is a necessary premise for the exercise of social power and where environmental costs are unequally borne by different groups. Like other species, humans are socially organised and this organisation profoundly affects the environment in which they live. Unlike other species, however, changes in the social organisation of humans are due to the complex interaction between two dimensions: the ideal and the material. Different ideas of what nature is and how it should be used generate different ways of organising production and reproduction and vice versa. Such historical, dialectic interaction between ideas of nature and social relationships is a major theme in the book's narrative[14].

Environmental change is a material process, but also a discourse, and environmental historians are engaged in understanding both. As a result, this book seeks to explain what environmental risk has to do with industrial capitalism and what culture underlies both.

PART I

Water and Revolution

Italian Landscape with Waterfall

In 1789, as the French Revolution was dismantling the old regime and disclosing unexpected scenarios in European history, the Swiss naturalist Ulysses Von Salis Marschlins made his way through the Liri Valley in the Italian Apennines and paused to admire the waterfall cascading to the right of the feudal castle dominating a town called Isola di Sora. He later recorded in his travel journal many of the features that would lend this valley its unique place as an exemplar of an iconic transformation in European power: from feudality to industrial capitalism via the enclosure of water. Von Salis' account of the place is thus a perfect vantage point from which to start our own journey into the Liri Valley's story: there we find outlined the contrasting elements of the landscape which would generate the great transformations of the years that followed.

For the most part, the valley that Von Salis evokes is a place of enchantment. Its core is the town of Isola, founded in ancient times on one of the numerous islands lining along the rivercourse. A few houses made of local stone and the campanile are grouped around the castle, erected on a cliff standing between two waterfalls; the urban space is surrounded by 'well cultivated hills' and deciduous woods. The Liri River completes 'the very beautiful scenery with the sinuosity of its course', while the countryside appears as one of those 'delightful places' whose beauty is 'rare even in Italy'[15].

As a traveller through late eighteenth century Italy, the Swiss visitor represented a literary archetype, that of the Grand Tour, by which numerous European writers and members of the educated middle class came to visit the *loci* of classical civilisations, contributing to shaping the myth of Italy as the 'garden of Europe'[16]. This genre interacted with, and continued, a local poetics of place: earlier literary descriptions of the Liri riverscape had depicted it as an evocative mix of natural force, beauty and history. Since the seventeenth century the language of the quasi-sublime had been employed to describe the river, as it came down from the mountains of Abruzzo to the territory of Isola, where, right behind the Ducal Palace, it split into two branches, one falling 'with very sweet roaring and great vapour', the other making a 'beautiful and very pleasing fall onto big rocks, resembling blocks of snow'[17]. By the early eighteenth century, the aesthetic appreciation of the waterfall had become mixed with a utilitarian vision of water and with the celebration of seigniorial control of the landscape: Isola was defined as the 'comfortable residence of and place [*stanza*] of delights to the Duke Boncompagni, who there enjoys a magnificent palace', endowed with 'waterfall, paper-hammers and other amusing and fruitful objects'. Even relish for the clarity and purity of the rivers' water was inextricably mixed with utilitarian appreciation of the wealth of fish – trout and carp in particular – that it fed[18]. A genuine enchantment was generated in all visitors by the sight of the waterfall, an expression of natural force and beauty: Isola was an 'astounding' destination for people wandering through Italy seeking waterfalls – in the words of one poet[19]; a 'work of nature indeed marvellous to the appreciative eye', which 'presents itself in a thousand shapes, all pleasing and surprising'[20] – in those of a late eighteenth century geographer. These descriptions of the valley were intended to celebrate both the wealth of nature and the existing social order. Indeed, the sight of the waterfall could not be detached from that of the feudal palace standing by its side and symbolising the sovereignty of local power over both the social and the natural world.

In 1793, a French artist named J. Joseph-Xavier Bidauld arrived in Isola and painted this scene with the title of 'Vue de l'île de Sora dans le royaume de Naples' ['View of the Isle of Sora in the Kingdom of Naples'] – now displayed in the neo-classical section of the Musée du Louvre (figure 1).

We can use this picture as an illustration of the Liri Valley described by Von Salis only a few years before. The picture itself, however, is part and parcel of the story. Not simply the immobile backdrop of a hypothetical theatre where the drama of history is played before our eyes, the elements of the landscape represented within its frame are the historical subjects of this book. First and foremost are the mountains: layered in an intersecting pattern, crossing with each other in succession from the highest peaks down to the plain which opens up in the lower centre part, these are the Monti Ernici – one among many mountain ranges forming the Italian Apennines and, in general, the physical structure of southern Italy. We will learn about the role the river had played in shaping them, about their soils, rocks and trees and how these became elements of the local agrarian environment; and about the socio-ecological links which humans had created between them and the valley, through what I will call the 'mountain-and-river system' of proto-industry.

The river itself, however, is barely visible in the painting. Though Bidauld did represent waterscapes, especially in his drawings and 'studies from nature'[21], he chose not to concentrate on the river in this particular case: his attention was especially attracted, instead, by the 'historical' environment (the castle and the buildings around it, seen from a distance that presents them as a unit). He seems particularly interested in the way in which this geo-historical unit –'The Isle of Sora in the Kingdom of Naples', as the artist named it – coalesced with the natural. The castle-town occupies the centre stage of the painting, located as it is at the intersection between the mountains, at the one point where the hidden presence

Italian Landscape with Waterfall

Figure 1. J. Joseph Xavier Bidauld, *Vue de l'île de Sora dans le royaume de Naples.*
Reproduced by permission of the Musée du Louvre, Paris.

of the river becomes visible in the form of a huge waterfall. The castle is indeed an immobile element of the landscape; yet, behind its permanence through time, there lies a substantial shift in its social role and meaning in the local space: from being a seat of 'total' power – political, economic, jurisdictional – to being one among several sites of mechanised industry.

The waterfall, for its part, is a key actor in the story narrated in this book. It is considered today the icon of the place, the most remarkable feature of the local landscape[22]. In Bidauld's painting, however, it does not occupy a very substantial portion of the represented space. It has a rather lateral position and inclination, adding an element of beauty and movement, of quasi-sublime, to the enchanted harmony of the place, suggested by other, more extensive elements: the sparse trees and 'unthreatening' woodlands[23] covering a good part of the scene, lying to either side as well as on the slopes surrounding the town; the suffused colours of the castle and other buildings, of the sky, lightly veiled with white clouds, and even of the mountains in the distance; and finally, the grace of the human and animal figures in the lower part, whose position and attitudes suggest a pastoral atmosphere. Taken as a whole, Bidauld's *view* was one of pastorality – of an enchanted relationship between humans and *their* environment. The figures in the picture are not drawn from an Arcadian mythology and put within the local space to give it symbolic meaning, but are intended instead to represent *real* local people (and animals) in their daily costumes and activities. They are marginal to the representation and do not dominate the scene, which is centred on the aesthetic harmony between the natural and the built environment. Yet their presence is not out of place – rather, it makes perfect sense and adds a sense of 'reality'. The chosen scenario is not one of wilderness devoid of human presence; the scene would not be complete without human inclusion.

The realistic intent of the painter should not mislead us into interpreting his representation as 'objective'. To begin with, the partiality of the representation lies in the very choice of perspective, which only allows us to see some characteristics of the landscape, especially its 'beauty', while hiding other, more prosaic ones. Perspective, above all, locates us – and even the humans within the frame – in the outside space of observation, where we can only enjoy the aesthetic value of the place while being safely detached from the material reality of life in the Liri Valley. Covered with woods, surrounded by hills, framed by distant mountains, this place seems totally devoid of both labour and property. As such, it resembles a kind of post-lapsarian Eden, where humans have managed to achieve a harmonious relationship with nature.

To understand these features of the image and how they influenced the unravelling of the story, we also need to take into consideration those elements that are outside the picture frame and yet essential to it. First, the artist himself and the reason for his painting that particular place at that particular moment. Bidauld can be considered a successful exponent of French neo-classical landscape art[24]. He had studied at the Dutch nature-painting school and his landscapes were usually vast panoramas *en plein air*, 'composed of harmoniously rhythmic lines' with the addition of 'picturesque and folk elements'[25], a type of composition that would become a popular genre for the entire nineteenth century. Known also as *Paysage d'Italie avec chute d'eau* ['Italian Landscape with Waterfall'], the painting was part of a broader group of landscapes that Bidauld completed during his long journey through Italy, upon a grant he obtained from the *Nation* in 1792 to 'palliate the lack of the usual clients, exiled or ruined under the Revolution' (as the Museum label informs us[26]). The artist spent much of his time in Italy in the countryside around the Liri Valley, the region of Abruzzo and the Roman Campagna, where he composed studies and drawings of rivers, mountains, hill villages and trees. He reached the Liri Valley after crossing the border with the Pontifical State and passing through the Pontine Marshes, a huge malarial floodplain south of Rome. He thus saw the sharply contrasting landscapes of the intensely inhabited and cultivated countryside of the Liri Valley and the marshlands just a few miles up north. His 'travelling gaze'[27] must have given him the sense of the Liri Valley as a place of appeasement between humans and nature, a place devoid of either wastelands or malaria, a healthy and joyful environment where humans had restored Eden through their past labour and could now proudly contemplate it as the place where they belonged.

Both the French Revolution and the connection between European artists and Italian landscapes[28] can thus help us make sense of the painting and of how the place was inescapably linked to broader historical and cultural scenarios. But other pieces should also be added to the picture, in order to understand its historical meaning. In these same years, new scientific instruments, such as modern cartography and especially statistics, were being forged, that, 'when allied with state power, would enable much of the reality they depicted to be remade', as James Scott convincingly argued in his *Seeing Like a State* (1998)[29]. Indeed, seeing both people and their environments through new lenses was an essential task of modern European statecraft in its effort at 'rationalizing and standardizing what was a social hieroglyph into a legible and administratively more convenient format'. The same effort at rationalising concerned the environmental hieroglyph.

At the time when Bidauld painted his view, in fact, the physical geography of the country had just started to be represented in topographical maps, which forever changed ways of seeing the land of southern Italy. The Liri Valley was included in the area called *Terra di Lavoro* (Land of Labour), a vast region formed of two distinct geo-morphologies: a plain area north and east of Naples, the celebrated *Campania Felix* of Pliny the Elder, occupying less than half of it, the other half being a mountainous and hilly inland area, profoundly marked by rivers and their valleys – one of which was the Liri. In those same years, 'modern' geography, the discipline that described national territories as object of both measurement and political-economic concerns, had entered the Kingdom of Naples. It was borne by a travelling philosopher, named Giuseppe Maria Galanti, whose descriptions of the country were read as a manifesto of anti-feudal politics. As we will see in Chapter 1, Galanti's geography introduced a political economy vision of *Terra di Lavoro*, but he also maintained a perception of its landscape as one of beauty, art and history. Other representations, more similar to the pictorial, were also produced in the same period in the form of local maps. Their coming into existence was part of a process of projecting the transformation of the Liri Valley into something different, more 'modern' and connected to the political economy of the State.

Some twenty years after Bidauld had visited the Liri Valley, when the first statistical view of the area was compiled at the will of a French King, the basic characteristics of the place were roughly the same as in the artist's *veduta*; what had changed substantially were the property and labour relationships imposed upon the landscape and its inhabitants.

Water and Revolution

A Road to Waterpower

Both Von Salis' 'enchanted view' and Bidauld's *veduta* of Isola Liri and its rural surroundings pictured the Liri Valley in the historical moment immediately preceding its 'great transformation' into the core of an industrial district. We can thus consider both visions as ideal epitaph inscriptions to the landscape of the old regime. A major change in cultural attitudes toward the natural environment – the Improvement project – brought about by Enlightenment ideas and politics all over Europe, profoundly affected the Liri Valley in the years immediately afterwards. Even Von Salis' apparently innocent representation of the place carried with it the germs of these new improving attitudes: he lamented the lack of a good road toward the capital, especially for the valley's woollen and paper products, and reported that some local worthies had mobilised on the issue, filing a petition to the King. An improving vision of the Liri Valley was being developed locally in the very same years by the clergyman Giacinto Pistilli, a native of the place, who was preparing a detailed report about the possibility of developing the area as an industrial district based on waterpower. That project implied the use of the Liri River as a waterway connecting the iron mines of Abruzzo – the north-eastern border region of the kingdom – to Naples, the capital and major port city of the country, where a new navy was being built. Requiring large, costly works, the channelling of the river was imagined as a unique opportunity to link transportation and power facilities, transforming the Liri Valley into an industrial district for the manufacturing of ironwork – especially guns.

The transformation of the Liri River into a waterway became the focus of repeated projects and debates in the course of the following century; yet, both geological and financial reasons conjured against the redesigning of the river into a straight navigable channel and this prevented the Liri Valley from assuming the conformation of other industrial river basins in the same period[30]. Nevertheless, the river did undergo major ecological changes, as its water was transformed into waterpower and the riverbed was enclosed into the dense patchwork of millraces, turbines and plugs that formed the essence of the factory system. For this to happen, however, a major shift in socio-ecological power had to occur – that from feudal to merchant control over water.

At the time of Von Salis's visit, such a shift was just being prepared. A map of the existing mill-sites in the valley was drawn in 1791, showing a landscape profoundly marked by the energy of the river and by the intense human use of this energy for production purposes (figure 2). This map follows the course of the Liri River upstream from the town of Sora, through its confluence with the Fibreno River, its circling of the town of Isola and its merging with the Gari River, where it took the name of Garigliano.

Though the Liri entered the valley with a fairly straight course, a few miles before reaching Sora it assumed a complex shape, revealing its interaction with the

A Road to Waterpower

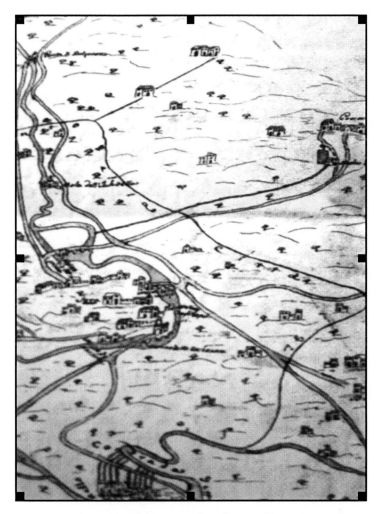

Figure 2. *Map of the water mills in the area of Sora*, 1791.
Archivio di Stato di Frosinone, Atti Demaniali, 65, 153. Courtesy of Ministero per i Beni
e le Attività Culturali (aut. N°11/13-03-2010), all rights reserved.

local space. The valley had a diverse geomorphology, with overlapping layers of different sedimentations – which the river itself had created at the end of the Ice Age and which gave the valley a particularly differentiated terrain. This geomorphology caused both the Liri and the Fibreno to split at several points into various branches, forming a number of little islands. In addition, and more importantly for humans, the diverse elevation of the place caused the two rivers to take a very tortuous course

Water and Revolution

and to form here and there falls of various sizes. This variegated landscape made by the mutual interaction of water and land, was further complicated by human presence. Entering this valley, the river encountered a much more densely inhabited environment than that upstream: the duchy of Sora comprised several minor towns and a number of villages and bordered the even more densely populated 'tenement' of Arpino. This complicates the task of discerning on the map the multiple courses of the two rivers and their branches from those of the pathways connecting the web of villages and towns in the valley. A number of bridges, pictured at the intersections of watercourses and roads, may be taken to signal the distinction between the two[31].

With its realistic picturing of buildings, country bridges, trees and mills, this map shows the local space as the result of a complex interaction between natural and social forces. One remarkable piece of information it gives us concerning human–water relationships at the time is the diffuse presence of mills and of a series of canals running symmetrically between two tracts of the Fibreno River, in the area called Carnello. Finally, though several road-lines crossed the valley, it is clear from their movements on the map that none of these could be regarded as a highway reaching the wider world. Advocated by many as an essential feature of the envisioned industrial landscape, that connection was just about to be created.

***

In a sense, Part I of this book recounts how a new road was built to connect the capital city of the Kingdom of Naples to a border province and its natural resources. It also tells the story of what followed: of the people, things and ideas travelling along that route for about one hundred years from the day it was completed, in 1797. The road to the Liri Valley carried with it two revolutions, the French and the Industrial. Throughout the nineteenth century, it carried soldiers and brigands with their equipment, civil servants with their correspondence, entrepreneurs with their machinery and tons of manufactured paper and textiles. When the Napoleonic wars were over and industrialisation burgeoning, another kind of travellers – Romantic writers – came to walk the road to the Valley: they looked at the landscape with the mindset of political economy and the aesthetics of their time and gave it a new narrative.

Following all these different people, the next three chapters will show how the Industrial Revolution was embodied in the Liri Valley.

Chapter One

The Landscape of Political Economy

You see plains covered with forests that with greater profit could be tilled, and mountains tilled that should be covered with woods. But if anyone were to introduce any useful novelty, he would soon be prosecuted by the tribunals, which are forced by the law to forbid any innovation.

G.M. Galanti, *Della descrizione geografica e politica delle Sicilie*

'You must forgive us, we are country girls... Except for church services, *tridui, novenae*, fieldworks, threshing, grape harvesting, servant flogging, incest, fire, hangings, army invasions, sacks, rapes, plagues, we haven't seen anything'.

I. Calvino, *Il cavaliere inesistente*[32]

This chapter looks at Political Economy through the landscape, the place where social and environmental changes were produced, experienced and embodied. It starts by describing the making of Political Economy as a discipline in the Neapolitan Enlightenment School, especially as regards the new connections being posited between nature and the nation[33]. Improvement – that is, the idea that nature had to be thoroughly redesigned in order to create wealth and strengthen the nation – was the core idea of classical Political Economy. As Carolyn Merchant writes, the Improvement concept was 'Capitalism's origin story as arising out of the state of nature through the evolution of private property'[34]. The Improvement project was grounded on two prerequisites: individual property rights and the State. The first would allow land reclamation and agrarian investments to be carried out by entrepreneurs whose private benefit was believed to coincide with that of the general public. The State, however, was the ultimate depository of the Improvement project: it would grant the acquisition and enforcement of property rights and would eventually become an improving agency itself where no available entrepreneurs were to be found.

Forming a new cultural attitude towards both nature and social relationships, the Improvement project was carried out by means of cartography and new modes of transport, as well as land enclosures and government edicts. Karl Polanyi described this process as a social and environmental revolution, 'Habitation versus

Improvement'. 'At the heart of the Industrial Revolution', he wrote in *The Great Transformation* (1944), 'there was an almost miraculous improvement in the tools of production, which was accompanied by a catastrophic dislocation of the lives of the common people'[35]. Agrarian enclosures since the sixteenth century and the factory system in the eighteenth had dislocated thousands out of their homes and transformed both their work and their relationship with nature into alienation. In Polanyi's description, the Habitation to which improvers brought so much havoc was a mixed socio-natural reality comprising the relationship of people with their environment as means of subsistence. Polanyi's aim was to contest the economic as the sole criterion by which to evaluate Improvement, by highlighting the dramatic changes 'facing a community which is in the throes of unregulated economic improvement'. Habitation was the life of rural people in its local dimension, and Improvement was what was disrupting it forever.

The Improvement project, however, did not exclusively carry negative connotations. In the European periphery, as in many late feudal areas of continental Europe, a much more positive, liberation emphasis was placed on Improvement as a means to free both people and nature from oppressive regimes of appropriation. More subtle, complex contradictions arise from the anti-feudal character of this political project and the material transformations it brought about in the land and lives of many people. To the Neapolitan philosophers and State reformers, for example, Improvement meant both the withdrawal of land and natural resources from the unproductive control of feudatories and the abolition of customary collective use rights. But these changes – the very substance of what reformers called 'political economy' – were inextricably linked to environmental change of a positive and much needed kind: they would allow marshes to be turned into arable land, the mountains to be reforested and people moved to the plains. In other words, political economy was to bring about a whole new, and supposedly better, order of ecological relations between population and resources, mountain and valley, land and water.

That said, the Improvement project did not proceed peacefully: not only did it lead to unprecedented dispossession of customary collective access to the environment but, in the Liri Valley, as throughout the old continent, it was carried out through sudden violent changes such as political revolution and counter-revolution, war and imperial dominion. Even when promoted by the enlightened absolutist State, the Improvement project assumed the character of a military campaign, war against nature and territorial conquest[36]. Overall, improving the land had as strong and irreversible implications for the people inhabiting it as for the non-human world.

Following the political and military events that led to the embodiment of the Improvement ideal in the Liri Valley between 1796 and 1806, this chapter will

A Road to Waterpower

try to make sense of how cultural, political and environmental change were related to each other on the European periphery in the Age of Revolution[37].

Figure 3. The Kingdom of Naples or the Two Sicilies.
From: Samuel Augustus Mitchell, *A New Universal Atlas* (Philadelphia, 1846).
Reproduced by permission of the David Rumsey Map Collection

The Landscape of Political Economy

Nature and Nation in the Kingdom of Naples

In the second half of the eighteenth century, the Kingdom of Naples was governed by a branch of the Bourbon dynasty that had gained independence from Spain in 1734 (figure 3). Located at the southern periphery of the old continent, the kingdom was nevertheless connected, through channels of intellectual correspondence and book circulation, to the outside world of European culture[38]. This intellectual connection was producing a new perception of the kingdom's position in the hierarchy of progress: that of a backward area[39]. The economist Antonio Genovesi, the most significant figure of Neapolitan Enlightenment, was the first to conceptualise the backwardness of the country within the theoretical framework of Political Economy. Like other classical economists, he considered himself first and foremost an educator and 'public happiness' the main concern of his discipline; consequently, all his works were conceived as parts of a discourse aimed at the Neapolitan elites on the moral necessity of progress and the best means to achieve it. He shared with other classical economists the social ideal of the landowner/producer as a 'rustic philosopher' pursuing the efficient manipulation of nature on international trade circuits and the vision of the countryside as 'a cohesive social unit dedicated to production'[40]. A simple catch-phrase – 'doing as in England' – was what Genovesi was wont to repeat as the core message of his teachings[41].

It was not only through the realm of ideas, however powerful these may have been, that the Kingdom of Naples was connected to the political economy of eighteenth century Europe. Genovesi's conceptualisation of backwardness found a very convincing expression in the structure of southern Italy's international trade relations. Due to the lack of a merchant navy, Neapolitan trade was dominated by British and, to a lesser extent, French and Dutch merchants who exchanged wheat and olive oil for imported manufactured goods worth more than double the exports. Although the imbalance was partially compensated for by more favourable trade with France, its position as a producer of agricultural goods with a low competitive advantage on Mediterranean trading circuits subjected the kingdom to continual commercial instability and discouraged long-term investments[42]. The peripheral condition within the shifting equilibriums of European (and world) trade made the economic reality of the country seem archaic and irrational, inciting the calls for change. Outside the realm of moral philosophy, though, the same peripherality created the material conditions for 'archaic', i.e. non-market, institutions also to seem perfectly rational, or, at least, acceptable[43].

This paradox of peripherality sets the context in which to look at the extremely intense social and political conflict that the kingdom experienced at the turn of the eighteenth century. In his most recent work on southern Italy in the Napoleonic era, historian John Davis states that the intensity and rapidity of political change occurring in the country from the 1790s to the 1810s was without parallel in the rest of Italy and Europe. To understand this, however, it is necessary to look at what had been happening beyond the capital city and its intellectual circles, but at the very core of Neapolitan society, institutions and culture: the feudal system. In the southern part of the peninsula, Davis writes:

> Feudal property was more extensive than in any other part of Italy or indeed western Europe. [...] Feudal landowners were entitled to raise levies over the local communities subject to their feudal jurisdiction. These rights and levies were countered, without being balanced, by the numerous collective use-rights that the local communities exercised on feudal land as well as on the extensive common lands that constituted one of the principal resources of the rural poor[44].

And yet, this world was experiencing strong forces of change, mostly originating in the unprecedented population growth in the latter half of the eighteenth century. Faced with increased demand for cereals and the possibility of amassing great wealth by commodifying access to the means of subsistence,

> the feudal landowners were often the first to violate feudal regulations by creating illegal enclosures, denying the local communities their collective use-rights or occupying sections of the common lands[45].

Something similar to what had happened in England over the course of two or more centuries had now occurred in the Kingdom of Naples in the space of a few decades, exacerbating social conflict and the crisis of political institutions. Following the great famine of 1763–64, the entire social and political organisation of the kingdom was undergoing extensive scrutiny and debate. The staggering number of deaths and the collapse of the wheat distribution system were interpreted as clear signs of the extent to which the economic institutions of the kingdom were dominated by corruption and inefficiency. It was in the aftermath of the famine, in fact, that Genovesi first published his *Lezioni di commercio* ['Lectures on Commerce', 1765], calling for radical change in the law and the replacement of feudal/mercantilist institutions with individual liberties and private property. From being theoretical principles, free trade and private property in land became the pillars of the necessary reconstruction of the Neapolitan economy and society. Reforming the State was no longer a matter of philosophical debate, but a moral imperative, a civilising mission and a project of 'national regeneration'.

In fact, after 1764 the Kingdom of Naples was probably the place, more than anywhere else in Europe, where political economy was seen as the necessity of reason. Abolishing monopolistic privileges to create free trade, eliminating feudalism in order to assert individual freedom and the full sovereignty of the State, dividing the commons to create private property in land, were seen as more than elements of a wise economic policy: they were the building blocks of a liberation project. Formulated by Genovesi, from the late 1760s onward this project had become the substance of every possible debate and of an emerging 'public opinion' among the educated elites in the provinces. When, in the 1780s, the clergyman Giacinto Pistilli prepared his project for the transformation of the Liri Valley into an industrial district producing iron for the nation – a project that implied the elimination of feudal control over waterpower and the channelling of the river – he was responding to this call for liberation and the improvement of the nation's natural resources.

This liberation project, it soon became clear, needed a reinterpretation of both the history and geography of the kingdom. It had to connect the Improvement ideal with the materiality of social and economic life in the provinces; in other words, it had to be grounded in science. Like their counterparts in other European countries, the Neapolitan philosophers were imbued with a far more comprehensive and un-specialised idea of science than ours. Philosophy was the unifying label under which mathematics and history, physics and metaphysics, natural history, ethics and philology were embraced as multiple aspects of reality. The emerging discipline of Political Economy was no exception: as an example of the common origins of economics and the natural sciences, Genovesi's teaching at the University of Naples was initially conducted under the label of *Commercio e meccanica* [Commerce and Mechanics], later becoming *Economia civile* (i.e. Political Economy). Created in 1754, Genovesi's was probably the first university chair in Europe under this label[46]. Like that of his contemporaries, Genovesi's idea of Economics drew heavily from one particular branch of the natural sciences, Physics[47]. With his *Discorso sul vero fine delle lettere e delle scienze* ['Discourse on the True Purpose of the Humanities and the Sciences', 1753], he popularised Newton's work in Naples, giving it a peculiar meaning as an instrument of empirical knowledge and social and economic transformation[48].

As this new Neapolitan scientific mindset encountered the modernising will of the Bourbon State, important new instruments of legibility upon the nation were created. First came a vast project of modern (i.e. mathematical) cartography, starting with a map of the capital city (pursued from 1750 to 1775) and progressively spreading throughout the kingdom, including the sea. The King also established a School of Topography in Naples (the *Officina Topografica*) whose first assignment was to compile a map of the royal game reserves in *Terra di Lavoro* (1784). In the following years, the *Officina Topografica* completed the measurement and triangulation of the entire country's territory and produced a colossal collection of maps, the *Atlante Terrestre*, published between 1788 and 1812. Commissioned by the State for military purposes, the Atlas expressed a new consciousness of the nation as a scientifically defined and numerically quantified entity, for the first time entirely 'visible' on paper[49].

Mapping the kingdom was just one of the many ways in which the Bourbon State participated in the process of 'time and space compression', which cultural geographer David Harvey considers typical of the Enlightenment project[50]. Prospecting and communicating were two other, very important, ways. The above-mentioned project for the channelling of the Liri and its use as a means of connection between the iron ore mines in Abruzzo and the merchant navy in Naples, represents a perfect example of these articulations between the State and philosophers, the capital city and the provinces and between various strands of the Improvement project. Influenced by the initiatives taken in this field by their Austrian counterparts, the Bourbons had become increasingly interested in mineral prospecting. In 1789, right before the revolution exploded in France, the Crown announced a competition for natural scientists with the purpose of forming a work-group that would travel throughout Europe learning about minerals, mining technology and metallurgy. That tour was to last eight years: among the vicissitudes of revolution and the Napoleonic wars, the seven selected Neapolitan scientists traversed most of Europe and returned with a huge collection of minerals and knowledge, which formed the basis for the foundation of the Museum of Mineralogy in Naples[51].

In a crucial sense, the Neapolitan Enlightenment project formed the link between what were conceived of as two distinct and non-communicating entities: 'nature' and the 'nation'. Improvement was theorised as exploration, measurement and colonisation; in short, as the recovery of an internal colony. A reinterpretation of the country's history and geography had been opened by Genovesi in his 1765 treatise on trade, which devoted a chapter to 'The state and natural forces of the Kingdom of Naples as regards the arts and commerce'. According to Genovesi:

> That which is now called the Kingdom of Naples, embraces the most beautiful, the prettiest and most fertile lands of present Italy, that were already famous for their philosophers, for the excellence of their law and legislators, for their military power, for their wars, arts and trades[52].

The almost mythical beauty and fertility of the country's land – grounded as it was in the classical age (Magna Graecia, or the era of Greek colonisation) – was evoked to support a discourse on the need to raise agriculture from its state of mere subsistence and restore it to its previous commercial supremacy[53]. The entire Neapolitan Enlightenment discourse, it could be noted, was one of restoration, of return to an ancient glory, located in the classical epoch, before the Barbarian invasions ushered in the age of feudalism. Following a long series of foreign dominations, the history of the country ended with the happy time of the new Bourbon State, established in 1734 as an independent kingdom. The problem with which Genovesi and the other philosophers mostly struggled was that of a modern monarchy which still retained institutions and laws of the feudal period, an intolerable incongruence which caused the country's 'backwardness' relative to other European monarchies.

By contrast, the wealth of nature and goods with which the kingdom's land was believed to be endowed was evoked by the philosophers through continuous reference to classical sources:

> Not only does the soil of our provinces hold all the advantages that Xenophon praises of Attica, but many more that he certainly ignored. For the land consists

mostly of plains, it is rich, watered and fertilised by deep rivers and streams; the air temperature makes it suitable for every species of plant, seed, animal and other things: those necessary to subsistence, such as corn, rice and legumes, olive oil, apples and every sort of herb, sheep, goats, cows, horses, mules, donkeys, pigs; and those which form the luxury of nations, such as silk, very delicate wines, delightful fruit, every kind of game and birds, abundant fisheries of the seas and rivers, etc[54].

Read in the light of the great famine of 1763–64, and given a mere glimpse of the physical geography of the kingdom, the above passage appears the merely rhetorical premise for a more sophisticated discourse on nature–society relationships. The basic message was that poverty and backwardness were not to be ascribed to nature but to 'man', and not only to common man but to institutions and laws – implicitly, to government. Though lacking the 'rich minerals' of Attica, the Kingdom of Naples had 'quarries that each year, and with overflowing abundance, grow from the land'[55], namely its agriculture, aided by the favourable climate.

The political economy of the Neapolitan philosophers can be thus more properly understood as 'political ecology', as a theory of nature–politics relations. The image of the open quarry growing from the land illustrates the most widespread idea the Neapolitan philosophers shared about the provinces that formed the bulk of the country's 'natural' wealth. The main problem, it was believed, was that the State in its various articulations – the Crown and the Neapolitan aristocracy forming the greater part of the jurisdiction – had no idea of the country it was governing, both because of the weak control that the monarchy had on feudal domains and because of the lack of good means of communication. Consequently, the philosophers put themselves willingly at the service of the cause of learning about the nation: following Genovesi's 'call for useful knowledge' a generation of enlightened aristocrats and landowners started to document the social and economic conditions of the provinces. In the process, they also expressed their condemnation of what they considered the archaic economic institutions of the kingdom, which formed obstacles to private property and commerce.

Getting to know the provinces was an essential step of the Neapolitan Enlightenment program, one so important that a proper method had to be developed for this purpose. This task was borne by Giuseppe M. Galanti (1743–1806), a student of Genovesi's, who had, Galanti wrote in his memoirs, inspired him with that 'fierce keenness for science [which] decided my fate'[56]. The science that was especially decisive for him was Geography, to be precise that new kind of Geography from Germany, which introduced the use of statistics and emphasised the quantifiable aspects of the Earth's surface. Galanti applied this new approach to the political problem of knowing the territory of southern Italy. The result was his *Della descrizione geografica e politica delle Sicilie* ['Geographical and Political Description of the Two Sicilies'], to which he devoted many years of travelling, data collecting and revising, and which appeared in five volumes between 1786 and

1794. Although Galanti was commissioned to write his Geography by the King himself, he was at times subject to ostracism and censorship; the book was finally received in European circles as a pioneering work of Statistics – the emerging science of national compilation. More than with the physical aspects of the territory, in fact, the *Descrizione* was concerned with its political economy, conceived as the numerical measure of its wealth. The import–export balance and the state of foreign commerce, internal revenues, the size of the population and its distribution throughout the kingdom all comprised the chief subject matter of the *Descrizione*, following Galanti's premise that 'a good geography is a more important book to the State than is generally reputed'[57]. It was an instrument of legibility on the territory of the nation, what the reformers used to call the 'inner Tartar', or *terra incognita*; an instrument both materially and theoretically forged by the philosopher and offered to the prince. As Galanti wrote in an over-quoted passage of the *Descrizione*,

> we want to visit the fields and huts of the peasant; to see how he farms; to examine what he harvests, what he pays, what he suffers; to discover the origin of our miseries, and, as is wished, to recover them[58].

The gaze of the travelling philosopher was, first and foremost, patriotic. He wished to see the peasant's life and landscape for he aimed at the public good of his country, aspiring to its improvement. History was part of this political project as well: in Galanti's vision, in fact, the love of the fatherland [*patria*] was a novelty for the country, because southern Italy had had no 'natural sovereign' since the thirteenth century, falling under the domination of foreign dynasties and 'a predatory and cruel government'. The primary aim of Galanti's work was thus to create the very substance of a new national consciousness, to be fuelled with 'the precious feelings of patriotism and the public good'[59]. With this aim in mind, both Geography and Statistics were useful instruments in order to build the science to which all human knowledge tended, namely Political Economy[60].

Though this approach was considered an expression of '*encyclopedisme*' and anti-religious views[61] and eventually drew the author toward censorship, the *Descrizione* was very influential in shaping a new consciousness of the relationship between nation and nature in the kingdom. Indeed, in the cultural and political climate of late eighteenth century Naples, the book's controversial nature was a great asset: it came to represent the vision of an enlightened bourgeoisie whose interests and opinions contrasted with those of the traditional elites. In the space of a few years, as we will see, this class found itself thrown into a 'Jacobin' revolution, soon after which the kingdom became part of the Napoleonic Empire. Although Galanti himself remained a royalist and was very critical of the revolution, the significance of his work for the making of a bourgeois consciousness had already transcended his beliefs in enlightened absolutism. As David Winspeare, one of the protagonists of the abolition of feudality in Napoleonic Naples, would write, the importance of Galanti's book lay in its 'examining, province after province, the abuses of feudality'. In sum, on the southern European periphery, Statistics had become a revolutionary science.

From the perspective of this book, the travelling gaze of the philosopher Galanti can be seen as imposing new 'disciplines' – Geography and Political Economy – upon the nation's body. This imposition raises the problem of inconsistency between expectations and the actual landscape, hence the need for improvement: Galanti looks at the provinces and has in mind both the classic texts and northern countries – European and Italian alike – to which he had travelled widely. More than anything, however, he thinks of Genovesi's idea of backwardness. The description thus becomes a discourse, full of rhetorical figures and myth. Galanti's gaze is an active part of a project for the building of a new nation. The real object of the philosopher's observation and concern is the State itself; the land and people of the provinces are only the mirror the philosopher raises before the King's eyes to show him the necessity of political economy[62].

In any case, Galanti's geographical description of the Two Sicilies gives a good idea of how – thanks also to the work of the *Officina Topografica* – the nature of the country was starting to be seen at the end of the eighteenth century. As in the *Atlante Terrestre*, so in Galanti's work, southern Italy is the world of the Apennines. His narrative focuses on the mountains, which, 'stretching the whole length of the country, encumber most of it in the north and west, which is called Abruzzo, and then branch all along their sides' until they come to the strait of Messina. The land of southern Italy is thus mostly upland, especially in its central part, whence it slopes towards the sea 'forming pleasant hills and fertile and delightful plains'[63]. Galanti's Apennines are limestone mountains with a granite base and their peaks start to be covered with snow at the end of October. To the author's eyes, however, mountain lands are not waste. Rather, they are an intensely inhabited and naturally diverse environment, whose trails are passable in the winter (except for the highest peaks in Abruzzo) and where different climate patterns create a diversity of soils and wildlife, as was described by the Greek geographer Strabo (58 BC–AD 25). Thanks to the mild temperatures, the vegetation in the plains is always green and different plants grow in different seasons. The sea winds mitigate the summer heat and 'one can see the spring when other regions suffer the hardest winter'. This is a land bearing 'crops of varieties unknown to the other countries of Europe' and the more the traveller approaches the country, 'the more he sees a land of a new and marvellous fertility and delight'[64].

Galanti's way of seeing southern Italy was largely influenced by classical sources[65]: along with the myth of fertility, the ancients had left stories of earthquakes, volcanic activity and the corrosive action of water in forming hills and valleys. As Galanti repeated, most of the soils were of volcanic origin, while most of the plains had been created by the rivers tearing down soil from the mountains and depositing it along the coasts, thus conquering land from the sea. Of the many rivers crossing the country, the author noted, all had lost the navigability that they enjoyed at the time of Strabo. This was a very important, indeed crucial, observation, one upon which Galanti himself, and many other authors after him, built their narrative of nature–politics relationships in the kingdom: the 'disorder of water'.

Despite their claims to building new scientific knowledge, Galanti and the Neapolitan philosophers could not help but rely on the authority of the classics as an almost indisputable source of information, especially when it came to the history of their own land. But they also actively and critically made their cultural traditions and heritage interact with the newer Enlightenment culture and political economy. Indeed, rather than scientific observation and measurement, Galanti's geography is based on a powerful mix of history, geography and Arcadian myth. 'This country of ours must have suffered terrible and extraordinary revolutions of nature', he notes, 'yet nature here is beneficial and this is the most beautiful country in Europe'.

It abounds in varied and useful products which 'open opportunities for its industry and commerce' – in other words, the country is filled with 'natural resources'. The interest of Galanti's political economy, however, lies in the way in which he represented nature–society relationships as a product of the country's history and politics. 'Not for lying under the earth's most fortunate sky or for the singularity of its nature is this kingdom worthy of the philosopher's attention', he claimed at the end of the first chapter, but instead for the great changes which men there have made'.

These man-made changes in the land of southern Italy had been disastrous: with the fall of Magna Graecia under the domination of Rome, beautiful cities, fertile lands and delightful places 'have been converted into deserts', while its inhabitants had become slaves. A general decline had taken place in both humans and nature: its causes had been wars, foreign invasions and political domination – indeed a whole history of ruin and 'reduction into barbarism'. From the barbarian tribes, which invaded the country after the fall of the Roman Empire, down to the last dynasty which ruled over the kingdom before its independence, all acted as conquerors and deprived the people of their status as citizens of their nation. This was the political cause of the current 'backwardness' of the country – as well as of environmental decline[66]. Indeed, beyond the emphasis of the first chapter on the natural wealth of the country, the *Descrizione* is filled with remarks on the disproportion between the population and condition of the country and 'our natural forces'. 'Where once there were cities famous for their population', Galanti writes in the chapter devoted to 'natural resources', 'today there are marshlands deprived of inhabitants'. And he adds, 'Moral causes concur with the physical towards our diminishment'[67].

With a few notable exceptions – including the area of Naples and part of *Terra di Lavoro* – the country seemed deserted. The Abruzzo region was populated by nomadic herders who in the winter left their home, together with thousands of animals, to go to the pastures of Apulia. A few of them went to the Roman Campagna to work the land and make charcoal, 'so they can pay their dues'. The people of the coastal marshes were fishermen and sailors, who fled the land, Galanti believed, because it was troublesome to them. There were generally few peasant farmers: the fields of Apulia were cultivated by the people of Abruzzo and Sannio; olives rotted in Salento at harvest time for want of labour. Even more scarce was the rural population of Calabria. The whole country represented a shameful contrast between the wealth of nature and the misery of people: a contrast which an enlightened government was called to overcome by means of political economy.

The very core of Galanti's discourse was thus the chapter devoted to agriculture, where he explained how the main obstacle to agricultural development lay in the kingdom's laws, by which '*the rural economy is not connected to political economy*' (emphasis added). The level at which this disconnection becomes clearer is that of property assets:

> If agriculture was once in an excellent state in these regions, it was because many indeed were those who owned [land] with free and exclusive property; and because this, as sacred, was granted and protected by the law.

The story used by the author to support his case was the same told by his teacher Genovesi almost thirty years earlier, now more firmly elaborated on the matter of property rights: the 'Barbarians', Galanti recalled, introduced 'the abuse of feudal government', by which 'it came in use to give acorns to one, wood to another, land to one, pasture to another' and so 'property was split and dismembered'. Political Economy was thus needed in order to bring reason to this irrational and barbarian order. This concept is exemplified in one of the *Descrizione*'s most famous passages, which can be read as the core of the Neapolitan reformers' idea of environmental politics:

> You see plains covered with forests that with greater profit could be tilled, and mountains tilled that should be covered with woods. But if anyone wishes to make any useful novelty, he would soon be prosecuted by the tribunals, which are forced by the law to forbid any innovation[68].

Here the moral imperative of political economy is expressed as an inextricable mix of ecological, economic and social imperatives: the frustration of the 'rustic philosopher' in seeing his attempts to rationalise agriculture impeded by the law, coping with the manifest irrationality of the agrarian landscape, gives the reader a vivid, quasi-pictorial idea of the necessity of improvement as a 'liberation project', and the status of private property as an objective, universal principle of rationality, representing the greater interests of the nation – rather than as one of several possible and contingent forms of socio-ecological relationships.

The discourse of contrasts between the (mythical) wealth of nature and the misery of politics is perfectly reflected in Galanti's description of *Terra di Lavoro*:

this once so favoured region, where Virgil saw vines 'husbanded' to holm oaks, olive trees, wheat and excellent pastures; where Pliny the Elder saw a land that could be sowed three times a year; and Polybius saw the most beautiful and famous cities of all Italy – this region had been ruined since the Roman invasion and the most notable evidence for its decline was in what Galanti called the 'disorder of water'. All its rivers, he reported, were navigable in the time of Magna Graecia, when people lived on the coastal plains near the estuaries and knew the art of river embankment. Cities and castles lay along these rivers in order to host the foreign merchants who sailed up them from the sea; one of these cities was Minturnae, at the mouth of the Liri: 'Nowadays nothing is there but a barge and a squalid inn', Galanti recalled. All that was left was the plentitude of fish – trout, carp, eel and nase – that the Liri offered, especially in the area of Sora[69].

The 'disorder of water' was thus caused by war and de-population; once the cities were destroyed, the waterworks were abandoned and water regained control over the plains. Finally 'the abuses of the feudal government' kept the watercourses from recovering their initial condition of navigability 'for the special interest of grinding and fulling mills was preferred to the general interest'.

Besides the disaster of the rivers, however, Galanti's *Terra di Lavoro* presented a varied and rich agrarian environment and was still 'the noblest and most fertile part of all the kingdom'. But it is hard even to try to see the country with the author's eyes; it rather seems as if he himself had described something from another book. That this is not entirely the case can be argued from documentation about the actual travelling Galanti incessantly undertook while working on the *Descrizione*. Nevertheless, given the state of the roads, brigandage and the abundance of unhealthy places, the author may well have limited his journey to some urban areas and immediate surroundings, deriving much of his information from conversations with local notables and visits to their bookshelves[70]. When it comes to the Liri Valley proper, for example, Galanti included it in the region of Montecassino: a land 'made of mounts, hills, slopes, valleys, plains, rocks and woods; yet, it does not lack its particular beauties', as he wrote, quoting Cicero – who was a native of Arpino. The author, however, was mostly attracted by the towns, all rich in antiquities, that lie in the region: of Sora, he briefly recorded the history since its foundation and the good trout and carp of the Fibreno.

In sum, to understand the *Descrizione* we should thus consider it as a testimony to the philosopher's gaze in his efforts to make a new State. For this purpose, the image in the mirror should be wisely composed: it should be frightening enough to move the Crown to accept the philosopher's advice and take action, but not so much so as to generate refusal and denial. Most of all, it is intended to inspire love of *patria* and zeal to improve it.

Improving the Valley

The line of anti-feudal reformism represented by Galanti offered the possibility of advancing a progressive political program while bypassing the virtually endless terrain of juridical controversy over feudal law. Whether or not grounded on legal titles, feudal tithes and customary practices had to be abolished as 'fetters' to economic efficiency and progress. This was an important theoretical step, for great efforts had been made by eighteenth century jurists to reinvent the feudal system, simplifying and rationalising customary–seigniorial rights, thereby seeking to overcome the main problem posed by feudal jurisdiction: its fundamental reliance on privilege and caste, rather than on the equality-before-the-law principle so crucial to Enlightenment culture[71]. The institution of private property, as Neapolitan philosopher Gaetano Filangieri – recalling Montesquieu – stated in his *Scienza della legislazione* ['Science of Law'], offered the possibility of grounding the legitimacy of the monarchy on new and more democratic principles[72].

The philosophers' writings might also be seen as reflections of the substantial increase in litigation between feudatories and local communities that, in the second half of the eighteenth century, resulted from the intensification of enclosures and other feudal usurpations, overburdening the tribunals of the capital city and paralysing the development of economic activities. But the judiciary's way-out-of-feudalism was clearly inadequate to produce any meaningful change in this state of affairs. While the jurisdictional aspects of the feudal system were agreed to be obsolete, in the economic sphere feudal landowners were increasingly adopting agrarian capitalist behaviour. This introduced the theoretical possibility that feudalism could be reformed from within[73]. It was no accident that the first and only reform that the Neapolitan *philosophes* managed to have approved was for the division of the commons [*demani*], mandated by royal edict in 1792[74]. The edict came ten years after the philosophers had been co-opted into the government, by the invitation to participate in a Council of Finance created in 1782. Like other eminent exponents of the reform movement, both Filangieri and Galanti served as counsellors, so that the Council 'came to embody the alliance between the prince and the philosophers'[75].

It was towards the end of this experience of collaboration that the duchy of Sora rose to the attention of the Crown. Two important changes resulted, with crucial consequences for the industrialisation of the Liri Valley: first, the construction began of a new road connecting Naples with the Abruzzo region through Sora; second, the duchy was 'de-feudalised', becoming a State possession. These things happened between 1792 and 1797; they were somehow representative of the highest degree up to which the reforms could be pushed in the kingdom and thus of the internal contradictions within the philosophers-prince coalition. Both initiatives were in fact primarily connected to the military concerns of the kingdom in a period of particular insecurity. The situation in France and the beginning of

Improving the Valley

Napoleon's campaign in Europe confirmed a longer-held preoccupation of the Neapolitan government with defence. But the road from Naples to Abruzzo was not justified merely by the need to strengthen borders; it received a strong impetus from the discussion of Pistilli's above-mentioned project, which in turn required a good means of transportation towards the capital, where the Neapolitan navy was being built. Given the technical difficulties of channelling the Liri, the only viable alternative was to construct a highway from Naples to the iron ore mines of Abruzzo, passing through the Liri Valley, where waterpower was available. Approved by the King, Pistilli's project was assigned to the Royal Army Academy.

It was in this context that the feudal possession of Sora and its territory first came to be disputed: the Duke of Sora strongly opposed Pistilli's project, claiming his feudal domain over the Liri and Fibreno rivers, whose water moved the fulling mills and hammers that formed a very substantial source of feudal revenue for the Duke. When, in 1795, Pistilli's project passed the Council of Finance's examination, the dispute with the Duke over water property had not yet been resolved. Nevertheless, nobody doubted that the State would ultimately acquire control over Sora and its waters: the question to be settled only concerned the amount to be paid[76].

The 'de-feudalisation' of the duchy of Sora is an extremely interesting case of transition from feudalism to capitalism because of the clear agency of the State and its military concerns. Behind the scenes, however, a crucial role was also played by the capitalists of the Liri Valley, whose voice was heard by the Crown through officer D. Cosmi, dispatched *in situ* on request of the merchant-manufacturers of Arpino. The latter saw the controversy between the Duke and the Crown as a unique opportunity to free themselves from the feudal monopoly over waterpower and finally gain access rights to the Liri and Fibreno rivers. Their aim was to acquire the full possession of water, thus accomplishing the transformation of merchant capital into industrial. This proved a rather complicated process, involving agencies and forces, like war, revolution and empire, that lay outside the control of local capitalists. The first step, however, was to withdraw the energy rent of the Liri and Fibreno rivers from baronial control and place it under the domain of a State whose economic policy was starting to be geared towards industrialisation.

When the King decided to acquire the Duchy of Sora from the Boncompagni family, the decision was thus based on the need for the State both to gain more secure control over border territories and to allow the development of private industry in the interests of the nation.

On a 1797 map of Isola Liri, we can now see a quasi-straight line running along the river, marked as '*Strada regia di Sora aperta [nel] 1796*' ['Royal Highway of Sora Opened [in] 1796'] (figure 4). Although that was probably only the enlargement of a pre-existing communication route (as may be deduced by comparison with the 1791 map of figure 2), this road was now part of a larger spatial connection between the border provinces of the kingdom – and their natural resources – and the capital. Even more than the road itself, what really mattered was the political geography surrounding it. By the time the new map signalled its existence – be it 'new' or otherwise – the Liri Valley tract of the road to Abruzzo no longer ran within a feudal tenement. It was by eliminating feudal control over waterpower, as well as by means of the new road, that capitalists in the Liri Valley could aspire to be part of a larger geo-political project: that of the development of national industry.

But events occurring in France, and spreading throughout Europe, did not allow the Bourbon State to implement its industrial vision. The execution of the French royals (especially of Queen Marie Antoniette, who was sister to the queen of Naples, Maria Carolina) and the beginning of the Terror, had already marked a sharp turn in the politics of the Neapolitan monarchy. This was especially bad news for the reformers: be they moderate or radical (the so-called Jacobins), they saw their position within the State apparatus and Neapolitan society dangerously compromised. After a Jacobin conspiracy was discovered in 1794, the situation

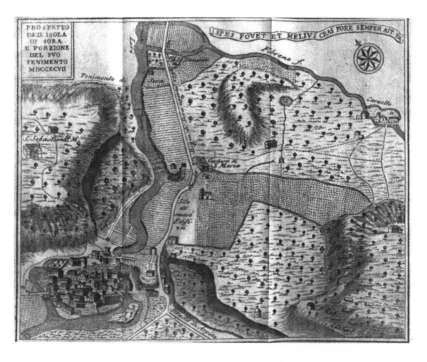

Figure 4. Map of Isola Liri and its hinterland, 1797.
From: Ferdinando Pistilli, *Descrizione Storico-Filologica delle Antiche, e Moderne Città e Castelli Esistenti accosto de' Fiumi Liri e Fibreno* (1824). Napoli: Stamperia Francese.

abruptly deteriorated. The works of Filangieri were banned, Galanti was deprived of funding for his statistical inquiry and was investigated for treason and other members of the Neapolitan reformers' movement were exiled, mostly to France[77].

In the aftermath of these political events, the Liri Valley found itself violently thrown into the Age of Revolution in its peripheral version. In the space of a few years, Sora's new road brought into the valley two military campaigns, two revolutions (one domestic and one foreign) and the Napoleonic empire.

Landscape and Violence

On December 28, 1798, 'walking along the road to Isola'[78], the French army first entered the Kingdom of Naples. This was a response to the offensive that King Ferdinand IV had launched against the French, who were occupying Rome, which resulted in utter defeat for the Neapolitan army. Following the defeat, the royal family and the court fled the capital aboard a British ship to Sicily. In early 1799 the Neapolitan army signed an armistice with the French. Nine days later, on January 21, 1799, the so-called 'Neapolitan Revolution' was proclaimed and a republican government took office in the capital[79].

The Republic only lasted six months, during which the Jacobins widely debated the extent to which the abolition of feudalism should be pursued and by what means. Intrinsically related to a military occupation, the Neapolitan Revolution was a very remarkable experience in the history of the kingdom: the contradictions inscribed within this unfortunate experiment in political modernisation have long intrigued historians and writers. The provisional government was led by Carlo Lauberg, a Jacobin pharmacist who embodied the link between science and revolution in late eighteenth century Naples: from 1792 on, he directed an Academy of Chemistry, which he himself had founded with the purpose of introducing Lavoisier's work to the capital, attended by most of the future Jacobins and members of the revolutionary government[80]. The poetess and journalist Eleonora Pimentel Fonseca (1752–1799) was among the most notable exponents of the Revolution, taking active part in government events and also leading the debate about the legislative reforms to be taken[81]. After an anti-feudal law had finally been approved in April, however, the royalists, led by Cardinal Ruffo, with the aid of British and Russian artillery, reached the capital in June and the Jacobins were forced to capitulate. They were then very harshly prosecuted: most were directly put to death and the remainder lost their property and were forced into exile.

The tragic end of the Neapolitan republic had two notable consequences: first, a period of 'royal terror' and harsh political repression started; second, the Jacobins began to be seen as heroes of a failed, but still necessary and just revolution, the failure of which was later interpreted as a clear sign of the backwardness of southern Italy, causing it to lag behind on the road to both political and economic modernity.

Indeed, from the very moment of its fall the revolution spawned its own narrative and a specialised historiography whose first exponent was Vincenzo Cuoco, a former member of the republican government. Cuoco defined the Neapolitan revolution as 'passive', remarking how it had represented the political vision of a social minority, while being sustained by a foreign army and defeated by popular participation in the counter-revolutionary attack. Cuoco also believed that the most important reason for the lack of popularity of the republican government lay in the feudal question, more notably in popular resistance to the end of the moral economy[82].

On its way to Naples, in fact, as well as within the capital, the French army found armed resistance. Even after the new government had been established, organised violence continued to dominate some provinces for months, even years, to come. One of these was the Liri Valley: as part of a border territory, it was beset by bands of royalist irregulars led by one Michele Pezza, called Fra' Diavolo, who operated in the area between the Tyrrhenian coast and the Liri River; inland, however, it was the band led by Gaetano Mammone that had been able to stop the French repeatedly by controlling access to the recently built highway to Naples. A former millwright, who used to lease the Duke's mills (according to local legends, he had become rich by finding a treasure in one of them), in the aftermath of the French invasion Mammone was able to quickly put together an armed gang, which he then used with the support of local seigniors both to sack the valley and to confront the French army. As a consequence of his actions, slaughter and devastation spread through the Liri Valley. On the morning of May 12, 1799, still mourned as one of the most tragic moments in the history of the town, about 350 people were slaughtered by the French in Isola Liri, whence Mammone had organised armed resistance; on his orders, all the bridges except one had been cut, so when the soldiers were able to force the gate and enter the town, most people sought safety in the local church, where they were easily found and killed. Those who tried to escape by crossing the river found the water high from recent rains and an overwhelming number of French soldiers scattered along the road. About 600–700 died in all[83].

The event marked the initiation of the relationship of local people with the French. As local writer Ferdinando Pistilli (a brother of the Giacinto mentioned above) remarked in his account of the 1799 events, the invading troops had smashed the rural idyll of the local landscape and its literary imagery:

> After two days [of sacking and fire] during which I lost two brothers and my house, the French left. This was, at the end of the past century, the fate of my homeland, which was once the object of beauty and enchantment, now of mourning and horror. Burning houses, ruined by violent fires, frightened the viewer; obstructed by relicts, the streets were barely passable, the crumbling walls dangerous and disastrous. For months such desolation turned away what remained of the citizens; and the women, who on the following Monday had a chance to flee the town, dispersed in various places[84].

Landscape and Violence

From the perspective of the Liri Valley people, who had seen death and misfortune advancing on them along the road to Abruzzo, the 1799 events were a clear manifestation of the violent face of the Enlightenment project, in both its monarchic and republican versions. Through wounded and raped bodies and destructive fury against the local environment, through demolished bridges, inundated fields, burned houses, the French Revolution had made its way into the Liri Valley. It had been branded on bodies and land[85].

The encounter between the Enlightenment project and violence in the Liri Valley was no 'military accident' on the road to modernisation: rather, the unfolding of that project as a violent process of transformation is part of what Nancy Peluso and Michael Watts have termed 'the political economy of access to and control over resources'[86]: accumulation processes – such as that leading to the industrialisation of the Liri valley – reshape the way communities distribute, reproduce and fight over their access rights to the local environment. 'All forms of political economy', Peluso and Watts maintain, 'have as their foundation the transformation of nature in social, historical and culturally informed ways'. Indeed, the perception of war as the pursuit of political economy by other means, to paraphrase Von Clausewitz's popular definition, can be dated to the Age of Revolution itself, in particular to the aftermath of the Napoleonic enterprise. The Improvement ideology that formed Enlightenment's ecological consciousness was intrinsically connected to the transformation of nature into economic resources to be made available to capital under the rule of law. Consequently, in its aim to acquire control over the Liri Valley, the Bourbon State came to be a key actor in the local process of capital accumulation. This would be poorly understood, though, outside the broader geo-political context in which the Kingdom of Naples was peripherally situated: that of absolutist Europe in its contradictory transition out of the 'old regime'. Reigning over one of the most feudal countries of the continental margins, the Bourbon dynasty was struggling to assert its territorial sovereignty against other forms of socio-ecological control. Meanwhile, the contradictions and unevenness of the European Enlightenment were exploding in the French Revolution and were then forcibly extended throughout the Napoleonic empire. Violence – in the Liri Valley as in many other places – was the way by which this broader pattern of accumulation became embodied in local spaces.

Six months is certainly too short a period to allow even a superficial evaluation of the environmental politics pursued by the Republic, or indeed any other aspect of its governmental life. Nonetheless, a glimpse into the events unfolding throughout southern Italy's forests in the period of the revolution can reveal much about the direction toward which that politics was pointing. In the conjuncture of war and violence which opened up with the 1799 revolution, the possibility unfolded for patterns of access to, and control over, land to be questioned through open revolt and counteraction[87]. This process mostly took the form of wood cutting and tillage, responding to the increased pressure that half a century of population growth and feudal enclosures had placed on the demand for land. The relationship of local communities with the Republic was often based on the attitude that the new State held towards these local insurgencies. Wherever the Republic took on a strenuous defence of property assets, it irreversibly lost mass consensus. Rather than the defence of private property, however, the Republic seemed preoccupied with the conservation of forests for military use. Several edicts forbade the cutting of trees suitable for ship-building in all the forests of the kingdom, public or private; penalties became increasingly harsh until woodcutting in any 'national forest' came to be considered as treason to be punished 'under the military code'. The Superintendent of the Department of Education (which had gained the forests previously owned by religious orders) was particularly active in denouncing the 'devastation' of woods under his jurisdiction and solicited the Police Committee to arm the keepers. The practice of branding trees reserved for the use of the Defence and Navy Ministry was also introduced in this period.

From the scant evidence we possess, the forestry politics of the Neapolitan Republic manifested a tendency to go against the moral economy of the southern Italian peasant, in the dual sense of enforcing privatisation on the one hand and State control on the other. No doubt much depended on the counteractions taken in different places by the peasants themselves, as well as on the stance taken by the local elites. Two trends can nevertheless be distinguished in the Republic's forestry policy: one is towards the reinforcement of State control over wood for the sake of defence; the other towards the abolition of common property. In fact, besides the military emergency with which the Republic had to deal during its whole life, the broader political programme it represented was the abolition of feudalism. This programme, as we have seen, was characterised by a noteworthy ambiguity and many uncertainties, also due to the greatly varied forms of access to land in southern Italian feudalism[88]. That complexity in itself can be seen as a major target for the improving–liberating mind of the Neapolitan Jacobins and as the greatest obstacle to the modernising aim of the revolutionary State, that of uniformity and simplification. One thing that was clearly seen as a problem to eliminate was the moral-economy aspect of the feudal system, namely that complex web of customary practices and legal titles around which the people of southern Italy organised their subsistence outside the sphere of the market. This is what Galanti had termed the 'rural economy', calling for its connection to political economy; this is what the 1799 protesters and woodcutters had been trying to assert, taking advantage of the revolution to restore their access to land against feudal usurpations and enclosures; but this is also what the Jacobins were unwilling to support, and what eventually decided their defeat. While they debated the law for the abolition of feudalism, the counter-revolutionary army of cardinal Ruffo, made up of royalist peasants and bandits, marched from Calabria towards the capital to restore the old regime.

But the restoration of the Bourbon dynasty did not stop the march, literally speaking, of history. A few years after the Bourbon return to the throne of Naples, in early February 1806, the French army once more crossed the northern border and walked the road to Sora. This time, it was an act of aggression directly aimed at the conquest and annexation of the country. Defeated by the Napoleonic army, the Bourbons had to leave the capital and Joseph – 'Giuseppe' – Bonaparte, Napoleon's brother, became the new King of Naples. Again, royalist bands and the Neapolitan army organised armed resistance and only in September could the French regain full control of the Liri Valley by forcing the walls of Sora, which was then sacked[89]. As for Isola Liri, all the inhabitants had fled the town for fear of massacre, leaving it in the hands of Fra' Diavolo and then of the French troops, who camped there for three months[90]. Again the bridges on the Liri and Fibreno were cut and this provoked the inundation of the countryside and the farms[91]. After things calmed down, the rich started to complain about the loss of their property, seeking compensation from the new State. The poor, on the other hand, had being doing all the work of repairing, rebuilding and replanting. Above all, they had been draining the land. In important ways, in fact, the destruction of the war had long-term effects on local space: one of the most serious problems that the French administration faced in the Liri Valley, as we will see in the next chapter, was the disastrous condition of the roads and the ruin of bridges, both related to the vicissitudes of war[92].

In the wake of the 1799–1806 events, Isola and Sora had been burned down and/or sacked twice; many people had been killed, raped, widowed or displaced; the local landscape had been devastated. And yet, in the above mentioned Pistilli's *Descrizione Storico-Filologica delle Antiche, e Moderne Città e Castelli Esistenti Accosto de' Fiumi Liri e Fibreno* ['Historical and Philological Description of the Ancient and Modern Towns and Castles along the Liri and Fibreno Rivers'], one finds expression of the inherent cultural contradictions by which the local elite made sense of the Age of Revolution. Though almost totally centred on the tale of military events, and complaining about local misfortunes, the book praised Isola as the most distinguished place in the Liri Valley, thanks to the industrial factories 'which have been introduced there', among which the author especially praised 'that of French-style woollen clothes belonging to Mr. Carlo Lambert, who has received the gold medal from the King'; and that of 'extra fine paper, that can compete with others in Europe, belonging to Mr. Pietro Coste from Lyonne'. This paper-mill – Pistilli went on – was located within 'a magnificent building'; and yet,

much more there shines the genius of the Frenchman, who was born to wonderful and exceptional things. He dared every effort and personal interest in order to give perfection to this Manufacture. And so much was he able to perfect it that it can compete with the best of France and England[93].

A new era had grown out of violence: individual rights, equality before the law and private property were its proclaimed cornerstones, but what really mattered to the author was the new place that his homeland had gained within the political economy of the regional space, the nation and the outside world. This new shining place was the result of the hard work and ingenuity of French industrial capitalists. Published in 1824, after the Bourbons had been restored to the throne, the book cannot be considered a paean to the French colonisers: it was instead a pure expression of the spirit of the times among the intellectual middle class of the European periphery.

***

This chapter has shown how the Liri Valley, a border province of the Kingdom of Naples, participated into the broader historical changes of the Age of Revolution. The making of a factory system in the valley was an effect of both new ideas about the country's economy and environment – the Improvement ideal – and of violent political changes. The Neapolitan philosopher of the late eighteenth century was motivated by the improving ideal and desire of transforming nature – land, wood, water, minerals, animals – into natural resources. Both Genovesi's and Galanti's works can be considered the expression of a 'recovery narrative'[94] in which society can be rescued from the fall from Eden (or Arcadia) by means of modern science and political economy. However, the Neapolitan philosopher's way of seeing the country in general, and the 'disorder of water' in particular, was neither fully mechanistic nor naturalistic: it was historical. Hydrological instability was rooted in political change, more than in the physical laws of nature. Nature was in no way a cause of decline and therefore the goal was not to dominate it but to restore a lost harmonic co-habitation with it, akin to when southern Italians were free citizens and owners. The very concept of the 'disorder of water' was a socio-natural hybrid: the order to which water was referred, in fact, was not that of natural forces such as gravity or friction (the basis of the movement of fluids) but that of society, of the social organisation of production and reproduction. In the philosophers' vision, nature's role was to sustain human habitation and commerce to the benefit of the nation: rivers were meant to maintain regular and steady flows, allowing cities to have harbours and flourish by exchanging food and manufactured goods with each others. Mountains were meant to be forested and plains tilled. People were meant to maintain this natural order of things – and they would, if only they were freed from barbarian irrational customs and allowed to gain and exercise private property in land.

The Neapolitan Improvement project – a peripheral version of European political economy – required the political liberation of nature, hence the emphasis on the abolition of feudalism. Although the Bourbon State was showing interest in an enlightened-absolutist version of the project, things rapidly took on a different shape and, before anybody could realise it, the Napoleonic empire arrived in southern Italy.

Chapter Two

Empire and the 'Disorder of Water'

Joseph Napoleon, by God's grace King of Naples and of Sicily, French Prince, Great Elector of the Empire: 'Feudality with all its attributions is abolished'.

(Naples, 2 August 1806)

We must regretfully admit that so great has been the negligence of our predecessors and our own in regard to water that for a long time now we have been subject to the deplorable state of suffering all the evils that can be expected from a bad economy of this substance either in abundance or in scarcity.

(T. Monticelli, *Memoria sull'Economia delle Acque da Ristabilirsi nel Regno di Napoli*, 1809)

This chapter narrates the changes in nature–society relationships that took place during the annexation of southern Italy to the Napoleonic Empire (1806–1815). It explains how, after the overthrow of the feudal regime by imperial rule, modernity came to southern Italy in the hybrid form of the colonial State. French rule was a peculiar experiment in modernisation, based on the Improvement ideal: a vision – shared by the colonising and the colonised elites – of the need to rationalise and develop the natural resources of the country by means of political economy. As a consequence, the Empire period resulted in an inextricable mix of contradictions: between the national, patriotic aspiration of philosophers and the peripheral-colonial position of the country; between the liberating aim of the French Revolution and its violent imposition upon land and people; between the ordering spirit of the modern/colonial State and the environmental 'disorder' it contributed to creating.

The changes in southern Italy's politics and law will thus be shown as they interlace with environmental change in the same period. Once the commotions of 1796–1806 were over and the long-programmed political changes were finally implemented, the colonisers were confronted with environmental devastation. River and land degradation, flooding and malaria became the preferred targets of a new generation of engaged Neapolitan writers, who felt compelled to reinterpret the environmental problems of southern Italy within the changed political context.

The chapter will thus analyse both the discourse and the material dimension of the 'disorder of water', trying to make sense of how river degradation and environmental vulnerability related to the new forms of political and environmental governance in Napoleonic Naples.

Liberating Nature

On August 2, 1806 the newly appointed King of Naples, Joseph Napoleon, abolished 'feudality with all its attributions' from the country; a few years later, with the establishment of a mechanised wool-mill in the former Ducal Palace of Isola, the industrial transformation of the Liri Valley started to take shape. To the colonising power, as well as to the educated middle class of the kingdom, the abolition of feudalism was an act of political liberation exercised on both peoples' bodies and nature: the mobility of labour and land as factors of production, in fact, must be considered among its most crucial effects. By 'land' the law meant, in physiocratic fashion, the bundle of economic resources that nature offered the nation. A crucial piece of the legislation, therefore, concerned the 'liberation' of water and this preoccupation was incorporated in articles 8 to 11 of the law for the abolition of feudality. This stated that 'rivers, any feudal right abolished, shall remain public property, and their use shall be regulated according to the Roman law'[95].

Once the 'barbarian' feudal order had finally been got rid of, an older order could be re-established: that of ancient Rome's water law. The latter accorded riparian landowners the right to use water without interfering with the course of the river nor with downstream landowners. Although clearly stating the 'public' character of water – which could not be appropriated, only used – such a system implicitly assumed the existence of private property over land and thus the idea that water use was a matter of private right. Restoring the Roman law in nineteenth century southern Italy was indeed a revolutionary act, for it implied a marked shift in property relations – from those centred on feudal tithes and customary communal access to those centred on exclusive individual property and written law. Although this transition had already been enacted for some decades, now the law sanctioned it, by adopting a revolutionary vision of what society-nature relations ought to look like.

In accordance with this vision, the 1806 law stated that,

> all the hydraulic machines belonging to grain-mills, olive-presses, fulling-mills, paper-hammers, ironworks, dyeing-works, copper-mills and similar – powered by public waters – shall be sealed as private properties; including the buildings, canals and other waterworks in service to the same machines.

Interestingly enough, the first preoccupation of the legislator was not with irrigation, as the agrarian character of southern Italy's economy would have suggested, but with waterpower. This fact might be interpreted in a dual sense: first, as a political economy vision on the future industrialisation of the country, based on the development of its hydraulic forces; second, as the manifestation of the status of the 'hydraulic machine' on the river as one of the most contested symbols of feudal power. The waterpower uses listed in the law text comprised virtually all the manufacturing activities controlled by seigniors and religious orders, who either made internal use of those machines, for the court or the monastery, or rented them out to contractors. As such, the watermill constituted a substantial source of feudal rent. People were forced to use the seigniors' machine, because they could not build their own, since waterpower use was restrained by the local lord. Late feudalism was, for the most part, the control over the energy rent of land and water[96]. The lack of market regulation in this sphere subordinated the price of goods to seigniorial needs and wishes –though regulated through a complex and locally varied set of customary rules. Clearly enough, the social group that most suffered from this state of affairs was that of local merchants, who found in it an insuperable limit to their need for expanding the sphere of control of capital. The law of 1806 did not, however, materially eliminate all existing hydraulic machines: rather, it transformed them into private industries with no special privilege on the use of waterpower, thus subjected to market laws and competition. Although the mills remained in the same hands and same places as they had been before, now they had to compete with other hydraulic machines which anyone was allowed to build on the same river. The principal aim of the law, in fact, was that of establishing an open access regime on water:

> In public rivers and on their banks – the law stated – shall anyone be allowed to build barges, bridges and any other work, upon obtaining from Us the license, which shall be given soon after seeing that [the above works] benefit the public and do not damage private rights.

The new regime of accumulation sanctioned by the 1806 law in the Kingdom of Naples is that of State capitalism in that special form of public–private balance that formed the dominant 'governmentality'[97] in Napoleonic Naples. The liberation of water from feudal right implied its full transformation into a natural resource to be used in the interest of the nation, as defined by the modern State. The symbolic power of the 1806 law was enormous and so were the transformations it enabled on the Liri and Fibreno Rivers. The power of water, hydraulic energy, became legally and materially available to capital. The socio-ecological conditions for the making of industrial capitalism in southern Italy had been put in place.

But before industrial capitalists could fully occupy the scene, a new set had to be erected on stage: that of the modern State. With a series of bills between 1806 and 1808, the new government established a 'centralised, bureaucratic administrative state along the lines that had emerged from the revolution in France'[98]. The Ministry of the Interior and the Council of State were the first institutions to be created on the French model; then a series of economic laws followed: abolishing pasturage rights in Apulia, reforming the old credit institutes in the capital and the revenue collection system, ordering the partitioning of the commons. During the same period the judicial system was completely redesigned, introducing the separation of powers, and the Napoleon Code became the new law of the country[99]. Among these first acts of the new rulers, the most important in socio-ecological terms was that of putting on sale huge amounts of public land: that acquired from suppressed religious orders and that belonging to the Crown. Along with 'the abolition of feudal tenures and the break-up of the baronial and communal *demani*' [i.e. the commons], John Davis wrote, this sale 'brought about a dramatic expansion of private properties, accompanied by the assertion of the exclusively private character of property and property rights'[100].

Guiding the political economy of the 'French decade' were two basic principles: uniformity of administration and private property. The two were strictly related: by ordering a new cadastre in 1809, unifying all direct revenues within the land-tax system, the newly appointed Minister of Finance, P.L. Roederer, aimed at a French-style fiscal and political reorganisation of the State, which could both produce and be sustained by the propertied class[101]. Roederer believed in the 'predestined secular rise of the middle classes and the replacement of land by capital'; he had been a protagonist of the French Revolution and was to become one of its first historians[102]. The programme of the Neapolitan philosophers had finally been put into practice by the French rulers.

The experience of French domination in the Kingdom of Naples can be thus considered a peculiar experiment in colonial modernisation. The building of the new State was a joint effort between the colonising and the colonised elites, who shared not only ideas of 'governmentality' and the Improvement project, but also actual government responsibilities. The exiled Neapolitan Jacobins were recalled home and given key positions in the reorganised State apparatus. Joseph Napoleon took his role very seriously: he travelled throughout the kingdom and gave a strong impetus to the work of the *Officina Topografica* (thirteen maps were published during his three-year reign)[103]. In the natural sciences as in politics, the French produced a number of the institutional reforms long advocated by the Neapolitan philosophers. First and foremost, the French put great efforts into the renovation of scientific institutions. The naturalist Giosuè Sangiovanni, an exile from the Neapolitan revolution to Paris, was recalled home in 1806 and granted the chair of Invertebrates Zoology at the University of Naples, where he introduced Lamarck's theory of evolution. In the same year the *Real Istituto d'Incoraggiamento alle Scienze Naturali* [Royal Institute for the Advancement of the Natural Sciences] was founded, with the task of supervising the application of scientific research to industry and agriculture[104]. In 1807 Joseph Napoleon also created the *Real Orto Botanico* [Royal Botanical Garden], annexing it to the Department of Mathematics, Physics and Natural Science at the University of Naples. This was the realisation of a project formulated in 1778 within the Neapolitan Academy of Sciences, which responded to the great emphasis that Linnaeus had put on botany, as 'a means to economic expansion'[105] and a State enterprise. With the institution of the *Orto Botanico*, 'bio-prospecting' was launched in the country as a far-reaching and organically preordained scientific project: the kingdom was subdivided into twelve sections (roughly coincident with the administrative provinces), fifteen official correspondents were nominated (mostly physicians and pharmacologists) and the first volume of a monumental catalogue of plants, the *Flora Napoletana*, was published in 1811[106].

The fundamentals of the colonisers–colonised relationship, historian Costanza D'Elia has argued, were grounded in a peculiar image of the nation's nature: in the cultural geography of Napoleonic Europe, redesigned as an aggregate of imperial provinces, southern Italy was the area most easily legible as an overseas colony, in the sense of a territory rich in natural resources waiting to be developed[107]. This way of seeing the country found support in the visions of nature–nation relationships that the Neapolitan philosophers themselves had been elaborating for several decades (as we have seen in the previous chapter) and formed the cultural possibility of collaboration. Like all colonial projects, however, the newly acquired territory of southern Italy was soon to reveal unexpected difficulties and obstacles: in short, its resistance against the improving mind of the French–Neapolitan rulers. One of the most clearly perceived and harshly fought against, was the 'disorder of water'. To deal with it, King Joaquin – *Gioacchino* – Murat (Napoleon's brother-in-law), who succeeded Joseph Napoleon in 1808, created what was to remain the kingdom's most enduring and effective instrument of environmental legibility/transformation, i.e. the *Scuola Reale di Ponti e Strade* [Royal School of Bridges and Roads]. Modelled on its French counterpart, the *École de Ponts et Chaussées*, and the first of its kind in Italy, the School was given the task of forming the kingdom's engineers on the basis of the most advanced physico-mathematical knowledge of the time. It was directed by the French General Campredon, who placed a special emphasis on hydraulics and mechanics, and was destined to have a leading role in the management of southern Italian rivers[108].

The Liri Valley was thoroughly involved in the above mentioned redesigning of the State. First, local political administration was reorganised into a new spatial hierarchy, centred on the district. Chosen by the French – against the pretensions of Arpino and Cassino[109] – to be the head-town of this new district, Sora hosted the *Sottintendente* [district governor], i.e. the terminal figure of the new chain of power that, for the first time in the history of the kingdom, organically connected the provinces with the capital[110]. A Tribunal of First Instance was also established in Sora.

Taking advantage of this opportunity, in September 1806 the newly elected *Sottintendente* Antonio Siciliani found his most compelling duty that of representing to the Crown the miserable state of the district, due to the devastations of war. Only two months before – he complained – the French troops had caused in Sora damages amounting to 3,000 ducats, 'and no doubt the population has lived in misery since'; hence he asked the State for 'some relief, among many suffered calamities', especially for the people of Isola, 'which is now in a pitiful state indeed'[111]. The Attorney General also wrote to the King, lamenting how, due to the destruction of the bridges, both agriculture and commerce were damaged[112]. What Sora and Isola deserved, in sum, was not special benevolent attention on the part of the prince – as the old regime might have predicated – but compensation for the 'repeated damage' which they had borne due to 'the position of the place'[113].

We will see how this early announcement of environmental degradation developed during the time of industrialisation; first, however, we need to turn to the new 'instrumental reason' of the imperial State and examine how the French rulers came to see the Liri Valley's landscape and people.

Seeing Like a Statistician

Probably the most compelling task of the new State apparatus – once this was established all over the country – was collecting information and putting it on paper, in the form of maps and statistics. King Murat, in fact, first ordered a survey of the 'uncultivated lands, lakes, ponds and marshes existing in the provinces'[114]; and then the compilation of a massive statistical investigation of the nation's territory, entrusting the direction of works to the Apulian economist Luca de Samuele Cagnazzi, who now occupied Genovesi's chair of Political Economy at the University of Naples[115]. The resulting *Statistica*, later called *Murattiana*, was a foundational moment of the French–Neapolitan colonial project. Preceded by other descriptions and cadastres, this survey differed, first of all, by its own definition: it was, in fact, the first to recall explicitly the use of that emerging science of the State, which was being developed with similar scope by other European monarchies in the same period. The most obvious point of reference, for the *Murattiana*, was the experience of the French *Bureau de Statistique de la République*: created in 1800, the bureau collected departmental memoirs written by Prefects in response to a questionnaire prepared by the Minister of the Interior. This effort of the revolutionary State at collecting information from its territory was part of what French scholar Alain Desrosières has called 'the process of creating equivalence'. The institution of the metric system, the unification of weights and measures, the imposing of a unified French language over local dialects, the declaration of universal 'rights of man', the abolition of privileges and guilds, the issuing of the Civil Code and the redesigning of the nation as an administrative entity subdivided according to Cartesian logic, were all pieces of a whole, spectacular effort to make the political principle of *egalité* effective on the land and people by scientific means[116].

Unlike the former tradition of surveys and descriptions, usually commissioned by the Crown and conceived as a 'mirror of the prince' for the use of enlightened absolutism, with its new scientific approach the *Statistica* offered to be 'a mirror of the nation': i.e. information of public interest, offered to the educated middle classes (and to the government which represented them) as an instrument of self-consciousness and rational administration. In the formation of the statistics, in fact, the most important role was that devolved to Prefects, namely the governors of districts[117]. Their writings, mostly addressed to local notables, among whom the Prefects were also counted, were initially conceived, as Desrosières writes, as a 'general descriptive discourse, easy to read and memorise, on the "wealth, forces and power of the empire"'[118]. By 1800, however, a quantitative approach was being superimposed on the narrative. The result was a hybrid, whose interest to the historian lies not in the reliability of the information it gives –in its instrumental ratio – but rather in its representing an ongoing process of cultural reconstruction of the nation. The revolution, in fact, involved 'changing not only the territory but also the words and the tools used to describe it'[119]. In the French, as in the Italian case, it would make little sense to look at these surveys as sources of 'objective' information, ready to be elaborated in numerical accounts: the interest of the survey is in fact in its showing things 'in the process of taking shape, before they solidify – even though they never grow entirely solid'[120].

The *Statistica Murattiana* was part of a similar agenda, although in its peripheral and imperial articulation. It made use of the French metric system, for example, but it did not rename the provinces according to mountains and rivers, as was the case with the French departments. With its imperial counterpart, however, the *Murattiana* shared two crucial features: first, it was launched in a time of precarious government conditions and war penury, which inevitably compromised the accuracy and reliability of the responses; second, it was informed by the desire to create a new kind of knowledge, suited to serve a new kind of political power. Cagnazzi, the man to whom Murat commissioned the direction of the survey, was considered a pioneer of the discipline in Italy: his *Elementi dell'arte statistica* ['Elements of the Art of Statistics'] was published in 1808–09; he was also the author of a treatise of Political Economy, published in 1813, which introduced Adam Smith's work to the Italian public. More than anything, Cagnazzi was a strong advocate of private property as 'the basis of every political economy'. Agrarian reform, in his vision, should consist in the division of the 'new lands' and 'the commons', without questioning the already established property assets, because the political economy of the new State had to be based on the sacredness of private property[121]. A crucial element of this vision thus became the need for the nation to gain 'new lands'; these could only come from two sources, partially overlapping with each other: reclaiming the marshes and enclosing the commons.

Neither, however, could be pursued in the district of Sora, as the report on the Liri Valley for the *Statistica* made clear. The Liri Valley that King Murat saw, through the eyes of surveyor Francesco Perrini, was already fully cultivated – from the deepest layer of available soil up to the highest possible level of mountain slope; nor were there any wetlands to drain. We shall now take a look at the Liri Valley landscape through the lens of the *Murattiana*.

The *Statistica* was the sum of single comprehensive reports from each province into which the national territory had been subdivided, while its narrative was divided into four broad groupings. The first concerned physical topography, including 'form and structure of the soil', hydrography, climate and 'spontaneous products': minerals, plants and wild animals. The second, and also the largest, concerned the 'subsistence and feeding of the population': crops and drinking water, the way food was processed and stored, local food habits, dress, housing and public health formed the bulk of the collected information, showing a primary ethnographic concern of the statistics with human behaviour and bodies. Public health was in fact a huge concern of the *Murattiana*, which had been initially conceived as a survey of medical doctors, nurses, practices of vaccination, medicinal plants and other remedies and healing practices locally available to the people of the kingdom. Third came what we might term the agro-ecosystem –'game, fishery and rural economy', in the terms of the *Statistica*: wild animals 'harmful to cattle grazing and agriculture'; fish – their local varieties and the multiple ways of catching them; cattle grazing and the main contractual forms related to it; then small domestic animals, bees and silkworms; then agriculture – the quantity of land and population, the division of ownership, the main crops, rotation and salaries of the peasants, the cultivation of corn, wheat and oat, that of hemp and flax, hay, orchards and 'perennial plants' (olive, vine and fruit trees); and finally, the measure of wastelands [*terre incolte,* or simply *incolto*]. Lastly, the fourth part of the survey was devoted to manufactures.

In the parts devoted to population and rural economy, the survey adopted a unified narrative of the whole province, giving only scant and occasional information about single places. Numerical tables on the distribution of population in the territory and classifications of animals and plants are mixed with comments and suggestions on the various agrarian practices or detailed accounts of methods of fertilisation, vividly exemplifying the hybridism of this early phase of Statistics. Consequently, aside from the topographical description, which is pretty accurate, the *Murattiana* cannot be used as a comprehensive source of information about any of the places that are mentioned in its narrative. The discrete method of investigation comes up again only in two other occasions – when it comes to wastelands and manufactures.

First and foremost, the statistician's gaze was topographical. It considered the land as a scientifically measurable territory, whose definition comes from both physical geography and administrative reason. The Liri Valley as such was never mentioned in the *Statistica*: its region coincided with that of the district of Sora, which in turn was part of the province (or department) of *Terra di Lavoro*. In the *Murattiana*, both history and myth disappeared and *Terra di Lavoro* lost its identity as a unique landscape, in order to be described, with an analytical mind, as an aggregation of discrete parts. 'To make sense of the statistical information', the surveyor claimed, 'it is advisable to cut the land that we shall cover in various parts and for each of them to investigate what is requested'. The design of the parts was dominated by geomorphology: like most of southern Italy, indeed, even the 'land of labour' was shaped by the Apennines, whose two main ranges crossed the province in different directions and degraded into hills towards the sea. The parts to which the surveyor referred, however, were not delimited by mountain ranges, but by the rivers dividing them; and here is where the Liri Valley enters the picture:

> The land lying between the Roveto Valley to the north and the straight bank of the Fibreno River to the south, irregularly cut through by the Liri River, and enclosed on every other side by the Apennines, is that of the district of Sora. Its shape is that of a little plain surrounded by mountains and by hills.

Except for the flat areas where Sora, Isola Liri and the village of Castelluccio were built, all that remained were mountain lands. Their appearance was deemed barren – their colour, in the distance, being 'that of the limestone: bold, and totally sterile'. Running for twenty-odd kilometres, these mountain ranges were mostly made of limestone, silica and clay; the highest peak was 1,720 metres above the sea level; they were covered with snow from November to April, sometimes May. 'The Apennines included in this space – the author commented – have the figure of enormous cliffs, inaccessible, overlapping with each other'. By contrast,

> on their slopes, and up to a certain point, the appearance is florid, for there exists a little cultivable soil and fruitful trees thrive. Almost all of the hills have a pleasant look, because they are for the most part completely covered with plants and cultivated.

The agrarian environment of nineteenth century Southern Italy was formed of two, broadly defined, landscapes, which historian Piero Bevilacqua has described as *terre del grano* [land of wheat] and *terre degli alberi* [land of trees].[122] The first corresponded to the arid inland areas characterized by extensive wheat cultivation; the second, to the mixed commercial farming of orchards and horticulture practiced around coastal cities and in Apennine valleys. The Liri Valley thus participated in the broader landscape of *Terra di Lavoro* as a 'land of trees'. Under their thin stratum (0.7 to 0.15 metre deep) of fertile soil, however, those hills were made of an almost two metre deep stratum of sand, under which rock and clay and sometimes sand again were layered[123]. This earth structure was a result of the action of natural forces, among which the river had a primary importance. The 1811 surveyor did not seem to be aware of this fact: it took roughly another sixty years for the naturalist Giustiniano Nicolucci, a native of Isola Liri and one of the first Italian anthropologists, to write about the agency that the river had exercised since the quaternary era in shaping the form of the land which humans inhabited and named the Liri Valley.

Seeing Like a Statistician

At the end of the ice age, Nicolucci wrote, the river ran at a level of roughly thirty metres above its present course, carrying a much greater volume of water down from higher mountain peaks and with stronger impetus. The combination of these forces produced a strong kinetic energy, with which the river carved its floodplain out of the Apennines; one important counterforce, however, was that of the large natural dams, formed of ice blocks, which stopped the watercourse in its rush towards the sea, forming large lakes. One of these depressions was that of the present plain of Sora. There the Liri slowed down and deposited the sediments which it had carved out of the rocks upstream. When the area finally dried out and the first humans came to live there, the valley was made of light alluvial soils layered on a compact stratum of travertine, scattered with sand and rocks. And so it remained when surveyor Perrini described the valley in the topography section of the *Statistica*[124].

The Liri that Perrini saw was still an impetuous watercourse, coming down from the mountains of Abruzzo thirty kilometres north-east of Sora, whose water fell 'precipitously' into the Roveto Valley, 'while running very fast between the Apennines on a rocky soil'. On entering the plain of Sora, just as it did in the quaternary, the river began a slower and milder course: it lapped along the town's east side and then, right at the doors of Isola Liri, it gathered the water of the tributary Fibreno. This was a minor course, which came out of various springs at the feet of the mountain called La Posta (just nine kilometres east of Sora): there, it formed a large and deep lake, where it was very easy 'to go hunting and fishing with rafts'. The waters of La Posta lake were so clear, that the bed appeared 'gemmed with lime-stones'. Next, the Fibreno passed beneath the ancient town of Arpino, lying on a hilltop, with a placid and regular course among the valleys; arriving at a place called Carnello, it fell from a high cliff and, 'after moving various machines, and especially those for soda-washing and purging the clothes which are worked in Arpino', it rushed towards the Liri and so entered the town of Isola, where it again split and formed the two 'beautiful and surprising' falls[125]. From there on the Liri resumed its rushing flow and, after gathering the waters of several tributaries, while 'carving out its course among the mountains of Sujo and Roccamonfina with impetus and great crashing', it took the name of Garigliano and finally went to form its delta in the floodplain of Minturno, fifteen kilometres from the Tyrrhenian Sea.

This topographical view of the land served to surveyor Perrini to introduce the discourse of the 'economy of water' – of human-river interactions. 'Such an impetuous and unstable river is not embanked', the surveyor remarked. Where the bed was large and the land plain, its natural banks were high enough to contain the ordinary floods; not so with the extraordinary ones, though, which usually inundated the countryside. The greatest damage, however, that the river caused to the fields resulted from the human practice of mill-damming: the system in use in the valley was that of planting temporary cane-brakes in the middle of the riverbed, so diverting the water towards the wheels during the summer months, when the flow of water was minimum. These palisades, called *cannizzi*, also blocked and collected the river sediments, so forming small islets which, in turn, caused the water to push against the natural banks, eroding the soil of the surrounding fields. The surveyor was pessimistic that a solution could be reached: only after 'an exact study on the nature of the soils, river embankments built at the right distance, and combined with the plantation of thick trees with sparse roots', could the problem be ameliorated.

When it came to irrigation, neither the Liri or the Fibreno appeared to be of any use. Their waters generally ran at a lower level than that of the fields and – in the case of the Fibreno – they were too cold; furthermore, to irrigate the plain of Sora without having stagnation or creating marshes would require complex and costly hydraulic works. The same was true for navigation: nowadays, the surveyor noted, large boats can go back up the river for only five kilometres from the coast, to a site called 'the stones', whence they took on board wheat, wine and olive oil. But the big rocks with which the riverbed was filled for the whole extension of its tributaries among the mountains, and the 'furious impetus' of the water, made it very hard to attempt navigation beyond that site. Only by clearing the riverbed of its rocks might navigation be extended further inland, as local chronicles claimed was possible in ancient times. In the years of the *Murattiana* survey, only small boats could be seen, some of which managed to enter the Rapido creek and get as far as San Germano (now Cassino).

At that time, however, plenty of fish populated the Liri. There were trout and shrimp in the river springs; barbells, squami anthias and round sardinellas as far as the confluence of the Rapido and from there the grey mullet, followed by the sturgeon near the river mouth. All along the river, people used to fish with hooks and bow-nets. Still richer in wildlife was the Fibreno, where there could also be found rovellas and large eels 'of very delicious taste'. Its lake and watershed were inhabited by wild ducks, shrikes and otters.

When it came to wastelands, the *Murattiana* became even more quantitative and detailed: it recorded the existence of approximately 10,000 acres of uncultivated land (*incolto*) in the district of Sora, roughly one quarter of the total surface; this notwithstanding, there was simply no land to reclaim as it consisted almost entirely of mountain pastures and rocks 'incapable of any culture', mostly located in the upper areas surrounding the valley[126]. If there was no land to reclaim on the mountains above Sora, there were no marshes in the plain to drain either – that is, not in the Liri Valley. Wetlands and swamps were all around: all over the coastal lowlands and, inland, the Pontine Marshes, just beyond the border with the Pontifical State. But, mostly because of its morphology, the Liri Valley was a rather dry land. Hence its eighteenth century iconography, partly reiterated in the nineteenth, which celebrated the amenity and pastorality of the landscape:

its significance, it should be noted, lies not in the typicality of that landscape but rather in its diversity from a broader context of wetlands, marshes and scarcely populated areas. And yet, the dryness and amenity of the land did not suffice to make its population happy. Intensely cultivated, the lands of the valley were owned by a mere ten per cent of the population[127]; to sustain themselves, peasants had to migrate seasonally towards the malarial plains of the Pontifical State, where labour was scarce and especially required. There they got sick – the author reported – 'with a mutagenic fever, which develops furiously and under which many die'[128]. Even when people did not work in unhealthy places, however, threats to public health could come from the river in the form of floods. These, Perrini wrote, had become a major threat for the towns and villages of the Liri Valley, since 'the deadly obsession with stripping the mountains there [had] become so ardent' and the waters rushed 'unchecked' towards the valley[129].

The next part of the *Statistica* where we can find detailed descriptions of the place is the section on manufacturing. Down in the Liri Valley, the surveyor observed, women did not make clothes, because they farmed the land. Most of them at least, namely the 'common people' [*volgo*], who worked along with men in planting and caring for the vines, 'husbanded' to elms arranged in double-row squares around the fields, as well as the olive and fruit trees. The spade was their tool, not the plough. The water needed for the trees was carried by hand from the river, traditionally a woman's task. This is how all the land in the plain of Sora and surrounding hills was cultivated and how the observers' gaze came to receive that pleasant orchard-looking impression of the landscape. It took a lot of work to create a 'land of trees' and women performed an essential part of it.

Women of the artisan class [*ceto*], by contrast, were kept busy weaving hemp cloth for the family in their spare time. The commercial value of this product could be calculated in ducats (the national currency) from prices at the local farmers' market, where sometimes the women brought their surplus[130]. The cost of whitening the cloths, which was done by dipping them into the river and then letting them dry in the sun, could be calculated too, even though only in local currency. These hemp cloths had a hierarchy of prices for the poor up to the 'comfortably off', depending on their quality and width. The rich, however, used fine linens imported from Naples. Planting hemp seeds in the countryside, the surveyor commented, was all that was needed in order to implement the manufacture of this ordinary cloth. Cotton was totally ignored in the area.

A different story was that of woollen cloth, where the term 'manufacture' assumes a proper market meaning. Traditionally, wool was a business of Arpino, the uphill town overlooking the plain of Sora, not far from where the Fibreno River ran before merging with the Liri. It had been so since the sixteenth century, 'being nowadays in an unfortunate decline', the surveyor remarked. Down in the valley, by contrast, wool manufacturing had been totally unknown until a few years before, when a few factories were established in Isola Liri, producing *castori* of fine quality and high price (which were only sold 'in the cities') and *peloncini* of medium quality and moderate price for the local middle class. The peasants and the poor still used rude cloths 'woven at home by their women and by themselves also dyed'. The raw wool came from the pastures of Abruzzi and Apulia. The Valley, the surveyor noted, enjoyed excellent fulling-mills: those of Carnello on the Fibreno River, which served the domestic manufactures of Arpino, and the new ones built in Isola along the Liri River, for the use of the new factories. These had introduced dyeing works as well; this task was formerly done in Arpino. Only men worked in these new factories, where they attended the hydraulic machines which carried out all the preparatory operations; since the newly built mills did not have yet spinning machines, the women spun at home, making slightly more than three pounds of thread per day. But this gendered division of labour was just about to change.

What the surveyor Perrini saw in the Liri Valley was a world in transition. Although it would take two more decades for the great transformation to be completed, its basic socio-ecological elements are visible to the historian. A revolution in the mode of production, based on the partial mechanisation of labour, had already caused the decline of proto-industry and was about to produce dislocation and change in gender relations. No protective institutions for the labourer such as guilds, labour statutes or mutual aid societies, were in sight, except for religious associations with no links to any art in particular: 'they give thus no aid to old or ill artisans, neither to their widows and children', the surveyor observed. On the bright side, he added: 'there is no talking about faults or abuses against the liberty and prosperity of the arts and manufactures'[131].

All this happened in a place where both land and water were being enclosed, by landowners upstream and by mill-owners downstream. Due to the end of the moral economy and to un-checked deforestation and milldams, the conflict of 'habitation' with 'improvement' had firmly taken place in the Liri Valley and it would grow much stronger as industrialisation proceeded.

Rivers and Revolution

It was in October 1813 that the *Sottintendente* Giuseppe Massone reported to the Minister of the Interior on the damage caused in the district by the 'disorder of water'[132]; this had several causes and only one of them (the ruin of the bridges) was related to the war. Two other major causes of inundations from the Liri were the five mills located on the river right upstream from the town of Sora and farming on the slopes in the Roveto Valley, further up in the mountains. The encumbering presence of the fallen bridges in the riverbed, as well as their striking absence from the urban landscape – which added to the already difficult state of communications – were enduring signs of the landscape of war[133]. Their destruction was the burden that the people of the Liri Valley suffered as a consequence of anti-French resistance; their repair was postponed until a very late date (1814) and with it the drying up of the roads[134]. But the other causes of inundation were related to a less contingent set of power relations, even more firmly inscribed into the landscape: that of property rights. As Massone described it to the Intendente of the province, the situation was extremely risky:

> Sir, I repute it very important to submit my observations about the Liri river-course in order to take into serious consideration the enormous damage that it will soon cause not only to the town of Sora, but to all the surrounding countryside, [throwing into misery] the nearby populations which draw their subsistence from it[135].

The tragedy of the Liri River, as told by the *Sottintendente*, started a few miles upstream, where the 'avid tenant farmers' [*coloni*] of the Roveto Valley had restricted the riverbed by planting willows and poplars along the banks, so much that the river now looked like a creek; the water ran at the level of the fields, and the road to Abruzzo 'in many places [did] not exist any more, being occupied by the river-course'. Environmental change was also occurring further downstream, where – curving around the walls of Sora – the river powered five mills: each of them used a large palisade between the riverbanks. 'The resistance they make to water in such a short space must naturally fill and raise the riverbed with sand and other materials' and this consequently induced the mill-owners to raise the palisades to a point where the water could only overflow the banks, causing the inundation of the town. 'Such a drawback', he continued, becoming stronger and stronger and threatening the destruction of Sora, had been felt over many years. It was then established that the mill-owners could not raise the palisades beyond a certain limit which was marked on little columns positioned near each mill. Some were still visible, but most were already well under mud and the riverbed was so high that the mills could not operate any more, 'remaining drowned under their own water'. Landowners and mill-owners alike were thus responsible for the damage, having caused the obstruction of the riverbed by different means, while 'competing with each other to get hold of the banks'. In many parts it was 'astonishing to see into how few feet of space the volume of a river like the Liri could be compressed'; in others, one could pass dry-shod from one bank to the other. Moreover, the great number of the willows tilting towards the water hid the view of the river itself.

In 1812, at the demand of a group of local landowners, the Minister of the Interior had authorised a project for the embankment of the Liri in the tract between Sora and Isola. These landowners were downstream victims of the 'disorder of water': their fields were inundated and the crops damaged due – they believed – to the mismanagement of others upstream[136]. But the problem, Massone observed, could not be solved with piecemeal defence works downstream. It was necessary to involve all the actors, and especially those responsible for the damage, in a wider effort towards restoration of the natural level of water – what he called the 'pristine' state. All agreed as to the necessity of this effort, but only in theory. 'Where you have many concurrent actors', Massone stated,

> it is no use to wait until each gets interested in the public good while sacrificing something of his own. The arm of the Government is needed, authorising a civil servant to identify the works and plan their execution and giving local powers the faculty of dividing the expense, while attending to the completion of the works. Otherwise, in a few years both the town of Sora and its tenement will be under water in the winter with the total ruin of the countryside and of human health.

This story of pre-announced disaster is totally based on human responsibility. There is no recognition, in Massone's words, of the role that nature can also play in both creating and destroying, of the power of the river itself. From the point of view of the *Sottintendente*, the agency of nature did not particularly matter. There was nothing he could do against it, since nature does not follow the rule of law. More interesting to him, instead, was the concurrence of social actors (individual owners, the State and its emanation – the civil servant) in forming the conditions of both disaster and its repair.

Although not a vague 'man vs. nature' narrative of declension, Massone's story is still extremely unspecific when it comes to the power relations and social responsibilities which underlie the foreseen disaster. He does not give us any clues as to who those *coloni* of the Roveto Valley were, who owned the lands they farmed, or why they were planting trees along the riverbanks. Neither does he say who the mill-owners were, or to whom the mills were rented. What becomes clear from his account, however, is how the 'disorder of water' was already a well established environmental question in the early nineteenth century Liri Valley – that is, immediately before industrialisation began to change the waterscape and local social relationships.

But when exactly did this 'disorder of water' begin to take its course? How can we make sense of Massone's story within the broader plot of socio-environmental change in the rest of the country?

To answer this question we need to turn to a second generation of writers who were protagonists of the Improvement project in southern Italy in its imperial phase (the French decade). In launching the alarm over the 'disorder of water' from the district of Sora, in fact, civil servant Massone was neither alone nor unheard. His voice actually resonated with those of many others in the early years of the French rule, all forming a new consciousness of environmental instability in southern Italy as related to the division of the commons and deforestation. According to historian Walter Palmieri, geo-hydraulic disorder had become a *Leitmotif* complaint, often lacking rigorous analysis, usually followed by a list of familiar arguments against peasants' ignorance, technical backwardness and extensive monoculture. A mix of problems whose interconnections with each other and with natural causes were often confused in a widely credited plot, whose most important function was that of orientating the legislative debate towards the methods and visions of modern 'silviculture', i.e. German-style scientific forestry. Beyond their internal differences, the writers of the Napoleonic period actually agreed on what they saw as a common undisputed enemy, the diversity of practices and use values that southern Italian 'rural economy' (i.e. the peasants) attributed to the forest. Be they pasturage or hunting or wood, hay and wild fruit collecting, all subsistence economies were invariably assumed to be destructive of the forest and responsible for environmental degradation. Goat grazing and other forms of agro-sylvo-pastoral economy (i.e. farming and grazing within the forest) were severely criticised as backward, pre-scientific practices. Under French rule, this general opinion took on a new political meaning for the Neapolitan reformers. Their common target could now openly be the old forest law established by the Bourbons in 1759, known as '*Incisione Arborum*', which was considered responsible for the intense spread of deforested areas in the previous decades. That law, a perfect mix of absolutism and moral economic vision, strictly forbade the clearing of forests for farming, while granting the continuation of subsistence use on the part of the peasants. Wood cutting was only allowed for the use of the navy. In the reformers' vision, by contrast, subsistence uses were to be eliminated while private owners had to be left free to dispose of woods since their interest would eventually converge with that of the public[137].

In the reformers' ecological consciousness, however, this vision coexisted with the political ecology discourse already set up by the previous generation of Neapolitan philosophers: i.e., a southern Italian version of the environmental declension narrative, in which nature and culture had fallen together from a classical golden age of harmony through the 'barbarian' age of feudalism, to end up in a present state of socio-environmental disorder. This discourse was articulated around the idea that the country's original, 'natural' wealth could be restored only by bringing back its pre-feudal institutions and especially the Roman Law with its absolute, inalienable individual rights. By introducing in the kingdom the Napoleonic Code, which was informed by a strong Romanist approach, the French had thus accomplished the first essential step of the restoration. The second step was that of recreating land tenure patterns similar to those which were believed to exist in the classical golden age, in a mythical pre-latifundium and pre-slavery society[138].

Such myth of a classical nature–society harmony is the most recurrent *Leitmotif* among agrarian writers of the French decade. Referring the current situation of the country to what the 'ancient historians' agreed to be 'the most fertile, the better cultivated, the most active in trade and the most populated' part of Italy, the exiled Jocobin Domenico Tupputi wrote in 1806 that 'one [was] tempted to repute their accounts as mendacious or exaggerating'[139]. The causes of decadence were all, the author remarked, in the 'errors of the legislation; carelessness of the Prince, laziness of the people; ignorance of both'. This picture was then contrasted with one of recovery and resurgence. Thanks to the new political regime, Tupputi wrote:

One sees agriculture flourishing under new laws and with agriculture even the arts
and commerce regenerating, population growing, uncultivated lands being tilled,
canals being opened, manufactures being created and towns being formed around
these manufactures[140].

By contrast, it was in the situation of the rivers that the 'negligence and
ignorance of government' had produced the damage most difficult to repair.

The rivers, which, conducted in man-made dikes, used to irrigate and fertilise the
land, today inundate it, forming sterile and deadly swamps. The harbors of Crotone
and Sibari, Salapia, Siponte and Canosa do not exist any more: the territory of these
once flourishing cities has become deadly for their inhabitants. [...] All the rivers
which the ships once navigated in search of merchandise from the internal regions
now have their beds covered with sand and mud [and], transformed in rash torrents,
devastate the lands that they once enriched[141].

Not only did these torrents, by their floods, 'take land away from agriculture' – the
author continued – but they also 'form[ed] swamps that infect[ed] the air', thus
causing the decline of rural population and the desertion of the countryside.

Most other authors of this period, even if primarily interested in discussing
agricultural practices, poverty and backwardness, considered it appropriate to begin
their discourse with what had become a widely accepted declensionist plot. As the
agronomist Filippo Rizzi, a member of the Neapolitan Institute for the Advancement of the Natural Sciences, wrote:

Razed lands, thick and unfruitful woods, uncultivated fields, unruly watercourses,
stagnant water and scrubs lie before the observer's eye. In vain has nature profligated
her gifts to our lands. Instead of profiting from them, we rebut them[142].

All writers invariably compared the present waste with the prosperity of the classical
age. 'I look at my homeland', Rizzi continued,

where the ancient Velia was, and I feel oppressed by shame. I search for the delightful forest of Trebazio and I find it thick with brambles. If Xenophanes, Parmenides,
Zenone, Leucippus, Alcidamante and Papirius were allowed to see their homeland
again, what sadness would surprise them in seeing it covered with wrecks? They
would not find the pleasant gardens, the leafy vineyards, the delicious fruit-trees
and other magnificence any more. They would be saddened to hear the hoarse
murmuring of frogs, which succeed in inhabiting the once cultivated and pleasant
fields, source of wealth and delight for the noblest Roman citizens. Now stinking
water stagnates in parts of the countryside, exhaling nitrogen gases which make
unhealthy that once so vivid atmosphere[143].

There are many elements in the discourse above which might form the
basis for a critique of the environmental consciousness of the agrarian elite in early
nineteenth century southern Italy. The theme of nostalgia and the classical remi-
niscences could be a starting point for the analysis of this *pastorale* and its meaning

Rivers and Revolution

in the context of country–city relationships. What is probably most interesting about the quotations above is the relevance accorded to the question of hydrological instability. Here is an important difference between the second and the first generation of Neapolitan reformers. While the rhetorical structure of the discourse on nature–society relationships is basically the same as in Genovesi and Galanti, it is the environment itself that has undergone major transformations in the last half century. In the first decade of the 1800s, in fact, the question of hydro-geological degradation is felt as much more impellent and its importance in the social ecology of the country would only increase in the following decades.

By 1810, the situation of Italy's rivers seemed serious enough to require a special explanation. An early Italian narrative of environmental declension, concerning water–land use relationships in the political and natural environment of the Mediterranean, began to take shape in that very period[144]. As regards southern Italy, the causal link between deforestation, soil erosion and the 'disorder of water' had begun to be recognised for its dramatic environmental effects by the 1790s[145]. This consciousness, however, would be fully developed during the French decade, in particular by the philosopher and mineralogist Teodoro Monticelli (1759–1845) – a secretary of the Neapolitan Academy of Science – who was the author of an 1809 *Memoria sull'economia delle acque da ristabilirsi nel regno di Napoli* ['Report on the Economy of Water to be Restored in the Kingdom of Naples']. His explanation for the 'disorder of water' is at the origins of an early southern Italian conservationism. Monticelli was one of three members of a committee on deforestation that the French formed soon after coming to the kingdom, with the task of delineating a project of forest law. A long debate followed, involving some of the brightest intellectuals and civil servants of the time, and the new law was approved in 1811; it also put in place a unified *Aministrazione Generale delle Acque e Foreste* [Department of Water and Forests].

A professor of Ethics and Chemistry, Monticelli can be considered a typical early nineteenth century 'naturalist'. His best known works concerned the minerals of Mount Vesuvius (the greatest part of his collection of lava and specimens was later acquired by the British Museum). In addition to being a mineralogist, however, Monticelli was also a philosopher, in the full sense of that term in eighteenth century Naples: he was a victim of the anti-Jacobin persecutions, spending several years in prison before the revolution, and then exiled to France. It is in this political context that his 'Report on the Economy of Water', first published in 1809 and reworked in several editions, can be properly understood[146]. 'We must regretfully admit', the Report states,

> that so great has been the negligence of our predecessors and our own in regard to water that for a long time now we have been subject to the deplorable state of suffering all the evils that can be expected from a bad economy of this substance either in abundance or in scarcity[147].

Monticelli's 'negligence' is that of governments and people alike, causing the depression of agriculture and grazing; and the 'long time' to which he refers is the twenty centuries separating the early 1800s from the 'very prosperous' times of the Magna Graecia. The fall from grace, as Genovesi and Galanti had stated, had started with the Roman conquest, which initiated a long history of wars, invasions and the devastation of environmental infrastructure – especially canals, dams and aqueducts – created by the Greeks. He conceded that the Romans took some care of water, but only when they were not busy with warfare. The real disaster, however, had begun with the collapse of the Roman Empire.

Despite his very long-term view of the 'disorder of water' – which he shared with others writers of the time – Monticelli clearly stated that the problem currently lay in the increased 'destruction of forests', which 'has inexplicably taken place for the last fifty years'[148]. The scarcity of wood and the filling of riverbeds with soil were clear signs 'of the foolish deforesting practiced until now and the impotence of the ancient laws in this respect'[149]. The most serious consequence of deforestation had been seen in the 'ruining of entire villages, taken away by the torrents, which grow before our eyes and acquire a devastating energy'. Despite all this, Monticelli complained, 'neither are we repairing the damage with new plantations, nor this obsession with deforesting has been stopped yet'.

Monticelli's explanation of the disorder of water, which became a 'scientific paradigm' for later generations, was indeed socio-ecological. He looked at the original cause of deforestation itself and found it in a circular process of vicious nature–society interaction. Deforestation, in fact, was due to the fact that most people in southern Italy lived on the mountains, because the plains were marshy and malarial; but destroying the woods and tilling the slopes to feed the population could only increase soil erosion and hydrological instability, which in turn increased the 'disorder of water' in the plain. This vicious circle, Monticelli believed, had gone ahead for centuries, since the fall of the Roman Empire. To stop it would require putting into action a whole, comprehensive water-and-forest scheme: draining and repopulating the plains, while reforesting the uplands.

With all their administrative efficiency and intellectual effort, neither the French nor their Neapolitan collaborators were capable of seeing the 'disorder of water' as the crucial ecological contradiction of the Improvement project and political economy. Their visions and explanations, on the contrary, rested on ideological assumptions that actually kept them from recognising the primary role that agrarian capitalism, gradually affirming itself in the last decades of the feudal regime, had played in the drama of deforestation and hydrological instability. Monticelli acknowledged that, in the previous fifty years, environmental change had been accelerating abruptly and the effects were clearly visible: rivers filled with soil, floods, malaria and landslides destroying entire villages. As for the other writers, however, for him the present was the time of socio-environmental recovery and resurgence, thanks to the French King, who, 'placing his glory in the happiness of his people and having understood the importance of repairing our water and restoring our woods', was already undertaking these works in several parts of the country[150]. Having destroyed all the old privileges and the barbarian laws that restrained the rich from investing their capital in agriculture and stock raising, Monticelli's story went, the nation could finally devote itself to the improvement of 'our very fertile lands' with much greater interest and cleverness than before.

What is particularly noteworthy here, however, is how the author connected the terms 'capital' and 'capitalists' with land improvement on the one hand and with the 'nation' on the other. A still clearer expression of this faith in the coincidence of private interest with the public good is in Vincenzo Cuoco, the Jacobin historian known for his interpretation of the Neapolitan 'passive' revolution. Towards the end of the French decade Cuoco wrote a report on *Rimboschimento e bonifiche* ['Reforestation and Reclamation'], where he drew a detailed prospect of the actions to be taken by the newly created Department of Water and Forests. His idea of how to prevent soil erosion and landslides was to create a mutual check system among landowners: each individual should be entitled to the legal means by which to keep his/her neighbour from enacting any reputedly risky novelty on his/her own land. By means of this system the free disposal of private properties would be somewhat limited, yet in a positive way: 'When the public good can be entrusted to the private interest, it cannot have a better guardian', Cuoco concluded[151]. However seemingly utopian, this practice of mutual checks among landowners would become the actual system of environmental control in the following decades. Explaining its failure to keep the land from the 'disorder of water' – as the Liri Valley story will sadly show – will consume much of the remainder of this book.

The landscape of pre-announced disaster seen by the *Sottintendente* Massone in the Liri Valley exemplifies what Polanyi called the conflict between 'habitation' and 'improvement': the ecological contradictions of an idea of the public good based on private interest. To make sense of these contradictions we need to consider a piece of legislation issued in September 1809 by the Minister of Justice, Giuseppe Zurlo (1759-1828) – a man whose views on southern Italy's political economy were deeply embedded within the actual process of change. Zurlo had been a student of Filangieri and Galanti; in 1789 he began a career as a magistrate in the highest courts of the kingdom, dealing with the fiscal, judicial and administrative aspects of the feudal question, especially regarding the common lands. Suspected of treason soon before the revolution of 1799, he retired from public life, but was then co-opted by the French as Councillor of State, then Minister of Justice and then of the Interior; he also served under the restored Bourbon State, from 1816 to 1820. Zurlo thus embodies the continuity of the Neapolitan Enlightenment project from the age of reform to that of revolution and then restoration[152]. His 1809 Memorandum on the use of public rivers[153] was to remain a landmark in the

legal history of water until the end of the kingdom, in 1860. It originated from
the Minister's preoccupation with enforcing the rule of the law for the abolition of
feudality enacted by Joseph Napoleon in 1806: as such, its targets were the former
barons and their legal pretensions over the now 'public' rivers and especially the
barons' attempt at conserving their former monopoly over the use of waterpower.

> Various complaints from town representatives and private citizens, who still experi-
> ence the old obstacles against the building of mills, have required that I took note
> of the causes which have delayed the enforcement of that part of the law for the
> abolition of feudality that destroyed the privileges, rendering free and common the
> use of water[154].

In the spirit of the 'liberation of water', the Minister thus ordered that –
with the exception of the navigable rivers, to be conserved in the interest of the
State – 'all other waters be left in their full freedom' [*libertà*] and all litigations
arising thereafter be resolved by the means of private law. Zurlo's emphasis on
the private character of water law went on to declare that the use of rivers be not
subjected to the granting of government permits, neither to 'any other restriction
than those of private law', which 'aims to direct their use and distribution among
those who have a right to participate and to grant the rights of property which can
be acquired on them'.

The Minister did, however, have a perception of the possible contradictions
between 'habitation' and 'improvement'; he stated that, wherever the interests of the
population were involved, either by use rights or by the possible damage to public
health, the use of watercourses should be subjected to 'regulations' [*regolamenti*],
i.e. formal agreements, issued by the town for: 1) partitioning irrigation rights and
2) checking on the 'regular course of water' by the means of public control over
hydraulic works and the water-flow. Both these types of regulation were justified
as pertaining to the superior interest of public health, since they would guarantee
the proper drainage of water and so the 'healthiness of the air'. The public was thus
allowed to interfere with the exercise of private property insofar as public health
was concerned. This regulation, however, was not mandatory: 'It shall always be
free [*libero*] to public authority to issue these regulations; yet where they do not
exist, the nature of waters', the memo repeated, 'shall not submit to any restriction
those who want to make use of them within the limits of the law'. To make sure
that this would be the case, the memo expressly forbade water-litigants from calling
for the intervention of public authorities.

As Chapter 4 will show, this contradictory and unfortunate piece of legislation was destined to give rise to a long history of litigation and environmental
disaster on southern Italian rivers.

***

Rivers and Revolution

This chapter has shown how, in the space of two decades of intense political change, between the age of reform and that of empire, the environment of southern Italy had also markedly changed. Floods and mountain slides of the early nineteenth century were not a discourse: they were very real and materially experienced phenomena, reported by a number of local sources, through those same agencies that the modern State had helped create and connect to the capital city by the means of national bureaucracy. By the time the French came to rule over southern Italy and the rationalisation of nature was finally put into practice, the socio-ecological contradictions of the Improvement project had clearly exploded. Especially through the massive sale of public lands – the former feudal and monastic estates – Genovesi's and Galanti's ideas had become part of a causal chain that eventually and unexpectedly reinforced the very materiality of the 'disorder of water'. This was not a locally circumscribed phenomenon, or even of merely national dimensions. Extensive deforestation has been documented in the nineteenth century Mediterranean[155] as one recent phase of a much longer process of alternating changes in forest cover, that began in the Neolithic after the introduction of agriculture through Northern Africa[156]. The Age of Revolution had taken place in a period of substantial reduction in forest cover all over Europe, now partly reversed.

Only a few decades earlier, however, in Genovesi's time, the 'disorder of water' had not been seen as such a widespread and serious threat; neither was it mentioned as part of the backwardness problem. The decadence of agriculture to which Genovesi referred was another thing than environmental decline. It is in Galanti's *Descrizione* and then in Monticelli's *Memoria* – that is, between 1790 and 1810 – that the 'disorder of water' becomes a major theme of political discourse in southern Italy. Crucially, rather than in recent processes of rapid political change allowing for the extensive spread of agrarian capitalism, these writers locate the source of environmental decline in a far away past, that of Barbarian invasions. They do so because the declensionist narrative serves a political aim: justifying anti-feudal reformism.

The political use of nature, however, was not invented by the Neapolitan philosophers of the late 1700s: it had its roots in the Greek tradition itself. The earliest reference to deforestation in the Mediterranean can be found in Plato's *Critias*. Here, as historian Joachim Radkau has recently commented, 'we hear about the good old days, which were already 9,000 years in the past, when there had still been much fertile soil and 'abundant timber' on the mountains of Attica'[157]. Yet Genovesi's and Galanti's classical reference is not Plato but his contemporary Xenophon, who had apparently a much brighter vision of the Greek environment of his days. Used uncritically – and selectively – as sources of undisputed truths, the classics served as mythical foundations for political discourses, based on references to the environment of the Magna Graecia. Archaeological evidence of the existence of malaria in southern Italy in pre-Roman times has recently undermined that paradigm of environmental declension[158]; moreover, and perhaps more importantly, it should be noted that the agrarian environment of Magna Graecia was a colonial one. The Greeks themselves were colonisers escaping from political unrest and socio-environmental problems in their country of origin; they found a 'new' land, southern Italy, already inhabited by other peoples, so imposed both their dominion and culture on it. According to Emilio Sereni's *History of the Italian Agricultural Landscape*, the Greek agrarian environment – based on the orthogonal plan for the distribution of suburban lands – was superimposed on the landscape of temporary clearings called *debbio* [slash and burn agriculture] of the previous inhabitants[159]. To the Neapolitan writers of the eighteenth and early nineteenth century, however, it was the Greeks, with their urban-colonial culture based on individual property in land – and not those primitive peoples – who should be considered the fathers of the *patria*[160].

The second generation of Neapolitan philosophers played a crucial role in shaping the way the imperial State saw the country. Their views were the product of the connection that the Neapolitan Enlightenment had predicated between the capital city and the provinces and between these and the European revolution. Their personal stories embodied this connection: the large majority of them came from the provinces; the places they described and the agrarian practices they commented upon were in their own homelands; in some cases, they themselves belonged to the class of agrarian landowners, a social minority in the feudal system, whose rise was nevertheless exactly what the whole Age of Revolution was about. In their participation in the Improvement project and then in the French government, they were implicitly asserting their legitimacy to speak for the nation and so the conceptual link between their visions and the public good. And their visions were strongly centred around the necessity of private property, invariably celebrated as the triumph of 'reason' over 'barbarism' and as the best means to liberate nature from irrational practices and to restore the pristine wealth of both nature and society.

This was the ideological context in which, by means of political revolution and imperial dominion, the rural elites gained access to land and water in southern Italy. We can now turn to look at the process of industrial transformation that took place in the Liri Valley soon after the liberation of rivers.

Chapter Three

The Ecology of Waterpower

> What 'satanic mill' ground men into masses? And what was the mechanism through which the old social tissue was destroyed and a new integration of man with nature was so unsuccessfully attempted?
>
> K. Polanyi, *The Great Transformation* (1944)

In the early nineteenth century, an Industrial Revolution took place in Italy's Liri Valley. Based on the mechanisation of woollen and paper production using hydraulic engines, it filled the Liri and Fibreno riverbanks with water-powered mills. The factory system was an 'ecological revolution': not only did it profoundly affect the prevailing mode of production, but it also reshaped gender and socio-ecological relationships – that is, the way local people made a living from the available resources within the property relationships that had emerged from the abolition of feudality. A new landscape – the industrial riverscape – resulted from the process and this in turn produced a new ecological consciousness, a new way of seeing society–nature relationships in the local space[161].

All these transformations taking place in the Liri Valley were a peculiar product of the Age of Revolution, during which both European political economy and technology underwent dramatic change. Technical innovations such as the Arkwright spinning machine and the 'endless' paper machine from Holland – both moved by waterpower – revolutionised the mode of production by enabling the mechanisation of most operations and an enormous growth in productivity. The ability of the new machines to push production to levels never seen before was at the root of contemporary perceptions of industrialisation as a revolution. The fact that the factory system also implied new social relations and a new landscape became clear very soon and contributed to confirming its revolutionary character[162].

This chapter will try to make sense of how the Industrial Revolution took shape and reworked both nature and society in the Liri Valley throughout the nineteenth century. It will describe the emergence of waterpower as the dominant energy system in the valley's industrial production and the consequent transformation of the Liri and Fibreno into densely industrialised riverscapes. It will narrate the mechanisation of labour from the point of view of workers and try to make sense of how the conditions of everyday life and family relationships were transformed within the ecology of the factory system. It will finally show how contemporaries narrated (and pictured) this transformation, thus elaborating a new consciousness of human–water relationships in the industrial era.

The Industrial Revolution did not, however, come to a pristine ecosystem – an enchanted place untouched by market forces or social change. Stemming from the political economy of the Enlightenment project, the factory system in fact took place in a landscape that was the historical product of long-term interactions between social and natural forces. As in many other European valleys and rural communities, industrialisation was preceded by what historians call the pre-industrial system of production, or proto-industry. To understand what the Industrial Revolution consisted of, we may thus begin by looking at the pre-existing system of socio-ecological relationships and how this was transformed.

The Making of an Industrial Riverscape

The power of water had long been known in the Liri Valley as a source of mechanical energy and had been appropriated by feudal lords and religious orders that were the primary holders of rights to land and water in the *ancien regime*[163]. In addition to supporting grain milling operations that dated back to the Roman Empire, by the end of the sixteenth century the energy of the Liri and Fibreno Rivers was used to move hammers and wheels that fulled cloth and pressed rags to process them into paper. These works were part of a pre-industrial mode of production based on manual labour and domestic industry. Yet pre-industrial work was also embedded within a changing landscape, in which capitalism was beginning to play an active part – the landscape of proto-industry. Indeed, by the time the French and the Industrial Revolutions had arrived in the Liri Valley, the latter had already been transformed into a new system of socio-ecological relations, linking mountain and valley, labour-power and water-power, feudal rent and capitalist profits. Meanwhile, the ecological consciousness of water was also changing. While common people saw rivers as part of the broader landscape of feudalism and moral economy, merchant-manufacturers started to see them in terms of waterpower, an economic resource to which they could assert rights without cost once the feudal system of river tenure was abolished[164].

Proto-industry is usually associated with the marginal agro-ecologies of hill and mountain communities with relative independence from manorial control and the need for additional income. Domestic production, especially of textiles such as wool and silk, enters the picture as an activity integrating agricultural labour, allowing 'upland communities' to survive and merchant capital to accumulate in the countryside[165]. This picture would be incomplete, though, without considering the work done by mountain rivers as they rushed down the proto-industrial valleys. After being woven at home, woollens were collected and taken to the river to be cleansed and compacted by means of water-powered pressing (or 'fulling') machines[166]. They were then returned uphill in order for the cloth to be manually transformed into finished products (garments, fabrics, etc). This mountain-and-river cycle of production reflected a socio-ecological division of labour: while merchants controlled the sphere of circulation and, partially, that of labour, they also needed the input of a natural power that, in feudal societies, lay beyond the control of capital: waterpower. As a consequence, proto-industrial merchants had to live with the feudal system and the feudal lords who were the masters of water.

Since the end of the sixteenth century, the Liri Valley had been part of a mountain-and-river system of proto-industrial production, whose core was the town of Arpino. Located on top of a hill on the western side of the Central Apennines, overlooking the Liri Valley, Arpino became part of the political territory of Sora in the early modern period. At that time, the 'arts' of wool and paper manufacturing were introduced by local lords, who profited from their feudal possession of the Liri and Fibreno rivers to install fulling machines, called *valche* (or *gualche*). Thus it was the feudal lord who initiated the cottage industry in the area, increasing feudal revenues by harnessing the labour-power of the inhabitants and linking it to his own possession of waterpower. Water and labour power together constituted an interdependent ecologic-economic system. This mountain-and-river proto-industrial economy of the Liri Valley, however, did not remain a static entity over the centuries. Capitalism gradually began to cut its way into it, as feudal power slowly eroded.

It was Duke Boncompagni himself who, in the early eighteenth century, nourished the rise of a class of local merchants by supplying them with large amounts of financial credit. This policy made sense within the logic of the feudal economy. The lord was a rent-seeker: he endeavoured to cultivate a group of merchant-manufacturers who were compelled to lease his *valche* and this increased tax revenues due to the growth of local incomes. In 1731, total feudal revenues from the town of Arpino amounted to 2595 ducats, while those from the *valche* of Carnello alone (located on the Fibreno River) were 3705 ducats. Thus emerged the political ecology of proto-industry, in which merchants were in charge of exploiting skilled labour uphill, while the lords exploited the mechanical power of water down in the valley[167].

After Sora came under Crown rule in 1796, both water rights and fulling mills were controlled by the Town Corporation, which rented them out to merchant-manufacturers. The monopoly over waterpower and the associated energy rent of the river had passed from the feudal power to the Crown and then to the Town Corporation[168]. The latter, however, was strongly influenced by the merchants, the most powerful economic group in town. These were the same people who had previously had to lease the fulling mills from the Duke. On the surface, very little had changed in the political ecology of the Liri basin –access to waterpower was still restrained through the mandatory usage of the existing fulling mills, now owned by the Town by virtue of a monetary transaction with the previous owner. Indeed, the economic policy of Neapolitan reformism had not gone so far as to 'liberalise' access to water. And yet the descent of proto-industrial capitalism from mountain to valley had started. Sensing the substantial change in Bourbon economic policy, merchant-manufacturers used every means at their disposal to accelerate the shift to a free access regime. An 1807 report from Town Councilor Luigi Guarnieri to the *Intendente* of the province exposed how the wool merchants in Sora had corrupted the town Supervisor, who allowed the income of the fulling mills to be diverted from maintenance and repair works. In other words, the merchants were planning to make the town's machines fall into ruin in order to gain permits to build their own. This strategy was evidently conceived to disrupt a public infrastructure that the local community had inherited from the State, in order to replace it with a private one. Doing so would allow merchants to cut the cost of manufacturing by acquiring full and exclusive control over waterpower. This would allow for the mechanisation of labour, which in turn would bring about a substantial increase in productivity under the 'mechanised' mode of production. Transformation of merchant into industrial capital was thus inextricably linked to a process of appropriating the river and to a political-economic vision of waterpower as 'free' energy.

Albeit ingenious, the merchants' corruption-obsolescence strategy was not revolutionary. It could take a long time to produce the desired result, which remained uncertain. Times were rapidly changing, however, and quick adaptation was required to keep pace with the impressive increase in productivity that mechanised industry could achieve elsewhere by means of waterpower[169]. For the merchants to become mill-owners and industrial entrepreneurs and for industrial capitalism to start up in the Liri Valley, a further step was needed: water had to be removed from the public domain and 'privatised'.

That further step was accomplished with the aid of the French. A decision in 1810 by the Supreme Court in charge of the implementation of Joseph Napoleon's law for the abolition of feudalism – the so-called Feudal Commission – granted the merchants of Sora their final victory over all other agencies in the control of waterpower. The decision finally allowed the manufacturers to build their own hydraulic machines and to power them with the waters of the Liri and Fibreno rivers[170]. Soon after, a report from the district of Sora, which was included within the *Statistica* of 1811, mentioned the existence of 'excellent fulling mills' on the Liri River, especially those recently built by the French entrepreneur Carlo Lambert for his new, 'great' *lanificio* [wool-mill]. This had been established inside the Ducal Palace, which the French administration had granted him in free concession for ten years, along with the palace's water rights and the title of 'Royal Manufacture'. The government also granted Monsieur Lambert the authorisation to import British machinery and continuous financial support through the following decade. In the space of a few years the industrialist was able to build several hydraulic machines so that, in 1818, more than 300 persons worked at his 'modern' *Lanificio*[171].

The Making of an Industrial Riverscape

The greatest novelty, however, had taken place right outside the town walls, where another French entrepreneur – Antoine Beranger – had established the first mechanised paper-mill in the district, the *Cartiera* of Santa Maria delle Forme. This mill was set up within the walls of a former monastery, following the suppression of some of the religious orders that had previously enjoyed the use of the waters to grind their grain. Again, the French administration had transferred the building, with its own water rights and infrastructure, to prospective entrepreneurs. The *Cartiera* derived its name from that of the monastery, which in turn incorporated the word *forma*, the Latin name for water duct. This was an 'English style' paper-mill, i.e. completely mechanised. Like the *Lanificio*, the *Cartiera* too had absorbed huge investments in fixed capital. Instead of the old water-hammers used in local paper manufacturing to press the damp rags contained within stone tanks, the factory installed the so-called 'endless' machine, made of toothed wheels and cylinders, which processed cloth into rolls of paper. 'This mechanism is entirely new within the county and the greatest results were expected from it', the *Statistica* had reported[172]. As the endless machine required a much greater quantity of waterpower than previous mechanisms, the mill was served by a system of canals from the Fibreno River[173]. By 1815, Monsieur Beranger had installed four endless machines, equalling the work of 48 former hammers. Energy was delivered by one large waterwheel, lifting water from the canal of Le Forme.

At the dawn of the second decade of the nineteenth century, the French and the Industrial Revolutions appeared in the local space as organic features of a unique and integral landscape – and, for now, were also embodied in the same persons. This new industrial riverscape, however, remained embedded within the old mountain-and-river system of proto-industry. Besides being the only example of mechanised labour in the area, the two French factories were at a very incipient stage. While Arpino's annual domestic wool output still amounted to 10,000 cloths per year, the new *Lanificio Lambert* had produced roughly 400 cloths a year in the period 1812–15. As for the *Cartiera*, it was suffering from technical difficulties due to the lack of skilled labour, such that it had been necessary to 'import' workers from England to fix the new machines.

It took some time to complete the industrial transformation of the Liri Valley. By 1820, however, changes in the use of the local space were clear enough to be seen by visitors and commentators: both the Liri and Fibreno rivers had been filled with waterworks, while new wool and paper factories occupied the space along the riverbanks. In addition, local capitalists had started their descent from mountain to valley. In 1816, the wool manufacturer Gioacchino Manna, from Arpino, had obtained from the restored King Ferdinand IV the concession of the former monastery of San Francesco with annexed water rights in order to build a mechanised wool-mill there. Two years later the factory produced 'excellent fine cloths' and obtained the title of 'Royal Manufacture', plus financial credit for 4,000 *ducats* and the permission to enlarge the existing waterworks. During the 1820s Manna, associated with Lambert, purchased part of the former Ducal Palace – that facing the lateral waterfall, called Valcatojo – in order to build a second wool-mill, with fifteen spinning machines and 45 workers[174]. The new woollen mill was valued at over 5,000 ducats: 860 for its new canal with two sluices, 1,000 for its two fulling machines, 250 for the two further fulling machines under construction, 800 for its hydraulic engine, 1,200 for two more engines under construction and 1,360 for the subterranean water duct running into the palace's basement[175]. At that time the Palace housed yet another wool factory, belonging to Giuseppe Polsinelli, with 44 machines connected to three hydraulic engines, plus three fulling mills. The factory also featured a machine-building workshop, 'working incessantly to supply this and the other local factories'. Another part of the palace had been occupied by a new paper mill belonging to Giuseppe Courrier, who installed the 'endless' machine there.

In the same years, Neapolitan entrepreneurs Lorenzo and Giuseppe Zino opened another *lanificio* in Carnello, near Sora: this was a huge installation, which combined under one roof the twelve ex-feudal *valche* visible in the 1791 map of Sora (figure 2), and whose big wheels had 'enough power to move many machines at once'[176]. In 1831 the factory had three hydraulic spinning machines with 120 spindles each, fourteen dressing machines, four gauzing machines and also featured its own iron-melting and cast-iron mills. It was estimated that the Zini had spent, in repairing works and in new machinery, roughly 115,000 ducats[177]. Until the early 1860s, this factory remained the biggest in the Liri Valley, employing 600 workers in 1853[178].

During the 1830s, the industries of the Liri Valley gained momentum. Due to a protectionist turn in 1824, aimed at encouraging national manufacturers against British competition, wool production witnessed a period of sustained growth and technical innovation. The four largest wool factories of Isola Liri alone (Manna, Lambert, Zino and Simoncelli) featured in all 125 hydraulic machines, suited to all phases of the production process. Mechanised weaving was introduced in some of the biggest mills, while many Arpino clothiers transferred their workshops down to the valley. Six of the 22 fulling mills recorded in Isola Liri in 1837 belonged to Arpino clothiers[179], while new arrivals continued to be registered in the following three decades. A mechanical industry developed alongside the wool and paper industries to build and maintain hydraulic engines and other machines in the local factories. An 1849 statistical enquiry on hydraulic engines on the Liri River registered fifteen mechanised factories between Isola Liri and Sora; more than 1,400 people worked in the five biggest *lanifici* alone. In the 1840s, the output of woollen clothes from the district represented 80 per cent of the kingdom's domestic consumption. Furthermore, eight mechanised paper factories had been added to the

The Making of an Industrial Riverscape

Cartiera delle Forme, whose paper would eventually reach the core of the industrial economy – the British market – to be printed on at the press of the London *Times*.

From the limited information available about the technical characteristics of the Liri Valley factories (and from 'archaeological' remains within Isola Liri's 'fluvial park'), the vertical waterwheel appears to have dominated the energy scene. With its ability to develop noteworthy amounts of power – compared to the horizontal wheel mostly used in grind-mills – the vertical waterwheel is considered an undisputed sign of industrialisation and 'the mainstay of European power technology [...] well into the nineteenth century'[180]. The waterwheel was the core of the new socio-ecological system, waterpower. In fact, it was part of a larger power 'organism' that extended from the river well into the factory, formed by millponds and reservoirs, millraces, canals, sluice-gates and all the mechanical components for the transmission of energy to the machines. This 'organic machine' – as Richard White would call it – inextricably linked the work of nature to that of humans, the operatives attending machines on the shop-floor[181].

The complexity of interdependencies in waterpower systems extended beyond the individual mill and entailed for the entrepreneurs some form of physical control over the environment around them. Inevitably, this involved other mills along the same stretch of the river and led to harsh litigation, a feature common to industrial basins elsewhere. In some cases, these problems led to various forms of coordinated command over the water-flow, either joint programs of upstream storage reservoirs and stream control[182] or (more rarely) the 'incorporation' of water rights on a large scale which allowed the creation of unified systems of waterpower distribution, such as that developed in Lowell, MA[183]. Speaking of the Liri Valley, it seems reasonable to argue that no such centralised energy-transmission system was available – not even for the Ducal palace, where several workshops coexisted – since each factory used its own canalisations from one or more points of the rivercourse. This web of individual water-plugs added to the already complex shape of the local watershed. One big waterwheel, usually located in the mill basement, was connected to the machines through a complex system of shafts and puddles transforming the rotating movement of the wheel into various forms of mechanical work[184]. Zino's woolmill, for example, used one 30 horsepower (hp) engine, connected to thirty different machines through an iron-and-leather transmission system. The average power of individual factories, however, was much lower – although it significantly increased in the 1850s: the fifteen workshops registered in the 1849 census featured one engine of around 6.6 horsepower each, while the three biggest workshops alone in 1865 had engines of 18.6 hp each. At that time, the number of registered factories in Isola Liri was thirteen and that of engines was twenty: some of the factories had installed more than one[185]. One explanation for these changes in power potential is that, by the mid-1850s, vertical wheels were in the process of being replaced by turbines – at least in some of the biggest workshops[186]. At the end of the century, 87 water engines were registered in the valley, distributed across 26 factories, producing a total power of 3286 hp. On average, each factory thus enjoyed 3.3 engines generating 124 hp in all, while the average power of each engine was 37.7 hp. In reality, however, substantial disparities existed between the factories: the old Cartiera delle Forme in Carnello achieved an overall 850 hp, the Cartiere Meridionali in Isola Liri 450; four more factories tapped total power of over 200 hp each, five over 100 hp, eight between twenty and seventy, while five only used 3–12 hp each. While the power potential of engines was generally close to the average (37.7 hp), a small group of users enjoyed both the greatest amount of waterpower installed in total and the greatest amount of hp per unit. At that time, the valley had a total industrial workforce (excluding grind-mills) of 2222 operatives, each endowed with an average power potential of 1.4 hp[187].

These sparse and unsystematic data do not allow the mechanisation process in the Liri Valley to be fully appreciated in quantitative terms. Nonetheless, they suggest an overall scenario in which the factory system grew both by expanding the number of mills and engines and the amount of horsepower along the rivers and by intensifying the power potential per unit-driver. This picture is fairly typical of early industrialisation in both Britain and New England, for the main characteristic of waterpower was its 'territoriality', its being a land-based energy system. Albeit an energy source whose output did not depend on agriculture (like wood, or muscle power), water was nevertheless inescapably linked to portions of the earth's surface – river basins – whose 'mobilisation' was very limited[188]. This made it substantially different from coal, which could be removed from underground and transported to factories, whose distance from extraction areas increased with the ever-improving efficiency of transportation. Furthermore, while only a small number of wheels and turbines could be built at a mill-site, there was virtually no limit to the number of steam engines that could be installed on the shop-floor. In waterpower systems, space was a very scarce resource. The best sites for developing waterpower were limited and competition over them was fierce, as we will see in Chapter 4.

First, however, let us turn to the labour side of the scene and investigate how mechanisation affected the life and work of people in the Liri Valley.

'I'll have your flesh at three cents per pound': Gender and Mechanisation

On the morning of 28 May 1852, a crowd of about 200 workers from the Polsinelli wool-mill gathered on the bank of the Liri River outside the factory walls. That same day, a new machine 'which by water-power sorts the wool, thereby replacing the labour of many arms' – a police report stated – had been shipped to Isola[189]. The fear of losing their jobs had driven the workers to throw the new machine into the river. This was the first (and only) case of a Luddite riot recorded in Isola Liri and it had to do with waterpower and gender relations.

'I'll have your flesh at three cents per pound': Gender and Mechanisation

Having been fired by the mill-owner, the workers decided to march towards the home of one Mr. Silvestri (a retired government official); crying 'Justice! Justice!' they called on the local notables to intercede on their behalf in order to be reinstated in their jobs. The Mayor, for his part, immediately called the *Sottintendente* who summoned the gendarmerie. At this point the mill-owner, as advised by the parish priest, tried to play down what had happened, claiming that it had been just a matter of female unrest: it was women, in fact, who were mostly employed in sorting wool. Deciding to investigate the affair, the *Sottintendente* discovered that 'some disgruntlement had arisen among the female workforce', since the young son of the mill's director – one Angelo Dephançon – had 'insulted them in their honor' by saying, 'in a short while the machine will come, your work will be useless and I will have your flesh at three cents per pound'. 'Compromised' by this offensive threat, the *Sottintendente* reported, the women had not hesitated to 'commit the act' with the aid of a few male co-workers. The latter, for their part, were also 'suffering a great deal' after hearing that power-looms would be shortly introduced in the district. They too would end up without jobs.

Aside from this issue, the *Sottintendente's* report assured, the workers were perfectly content with the treatment they received from the mill-owner, nor did they have anything to complain about in their jobs. By virtue of this version of the facts and their motives, the *Sottintendente* managed to reinstate all the workers in question, requiring only that a 'momentary punishment' be inflicted upon them, 'without taking away their bread' – also given that the pieces of the machine could be easily recovered from the riverbed. Once things had returned to normal, however, twelve men – the young Dephançon among them – were identified as being materially responsible for the riot and arrested 'as a reprimand and example to all'. Thanks to these measures, the *Sottintendente* could boast that he had prevented any further public unrest; most of all, he wrote, 'no idea, presumption, nor fact followed, that might even vaguely be considered of political relevance'.

The episode throws some light on the social reality of Isola Liri in the mid-nineteenth century. After three decades of industrialisation, social relationships still resembled those of a patriarchal rural society, where subordination of the workers to the existing order is unquestioned and the ruling class is legitimised by its ability to save traditional values such as the defence of feminine honour and the protection of jobs. The problem of the unequal and threatening relationship between labour and waterpower in the factory system was overlooked and transformed into one of preservation of the status quo: in exchange for the arrest of the offenders, the workers lost their opportunity to question the existing organisation of work, one in which industrialists were the owners of water and labour and could use both as they pleased. By exercising clemency and forgiveness, as well as moderate punishment, both the *Sottintendente* and the mill-owner, Polsinelli, reaffirmed their roles as guarantors of an inescapable socio-natural order.

The 1852 riot also throws some light on the place of women within the local social ecology: the episode must be read against the background of a long history of labour and gender relationships. In her detailed study of family and work relationships in the Liri Valley, historian M. Rosa Protasi has shown that mill-owners used a female (and child) workforce as a flexible, low-cost tool to offset cyclical market crises: when prices were high, they tended to preserve the traditional division of work, with women employed in manual preparatory operations (such as cloth selection), while men attended the hydraulic machines and the hand-looms. During depressive cycles, women and children took men's jobs in the factory, but maintained their previous salaries, which were substantially lower: this allowed industrialists to reduce production costs through additional exploitation of the workforce. In the long run, however, both the woollen and paper industries tended to dismiss female workers from the factory, especially after they had married and given birth to children. As many labour historians have recognised, this was a generalised effect of the late nineteenth century passing of national bills to defend the health of women and child workers[190].

These unequal gender relations of production help us understand, on the one hand, why the allusion to Polsinelli's female operatives' bodies as 'flesh on sale' sounded so realistic a trigger to women rioting; on the other, why the ruling elite could so easily dismiss the machinery question as one of paternalistic protection and defence of existing relations of production. This entrepreneurial behaviour was politically 'rational' in the sense that it responded to the fear of social ills related to the Industrial Revolution in England: maintaining the mountain-and-river system of complementarities between mechanised spinning and hand-loom weaving seemed the best possible way to avoid the manifestation in Italy of the English spectre.

This permanence of quasi-feudal relationships came at an extremely high cost for the local population. Socio-ecological relations in the valley had substantially changed since pre-Napoleonic times and there were few possibilities for surviving outside the factory system – except for out-migration. Though resembling a 'moral economy' scenario, the situation in the mid-nineteenth century was a 'liberalised' Liri Valley, where both land and water had been enclosed and mill-owners had succeeded in replacing the power of the feudal lord with their social power. The most striking example of this is the fact that, even in the 1870s, salaries were still commonly paid in food-crops: the industrialists being very often also local landowners who set the price of their products, this expedient granted them an additional source of profit and total control over the local economy. Industrialisation had created in the Liri Valley a new system of socio-ecological relationships: let us now try to understand how this system worked.

Improvement vs. Habitation

Creating a factory system out of a proto-industrial one was a costly process of socio-environmental transformation. Industrialists had to invest first in land ownership, then in extensive canalisation works and in new machinery, often imported from Britain and France; they also had to contribute to the creation of a local mechanical expertise, in order to maintain the machines and build new ones[191]. The appropriation of the river by mill-owners thus increased the productivity of water, capital and labour invested in it.

The power of water which industrialists exploited, however, had not been created by them: it was a 'free' gift of nature, transformed into a form of capital. Here lies, historically, the first ecological contradiction of industrial capitalism, namely its transforming both nature and labour into capital. In the new system, work was substantially transformed and so was the relationship between work and the natural world. In fact, one way of seeing the first Industrial Revolution is as the appropriation of a non-biological energy rent (waterpower) by private capitalists. Such appropriation allowed the mechanisation of manufacturing, integrating and then supplanting domestic labour.

Despite the shift being gradual and the two systems long complementing each other, what matters is the relevance of the inanimate, water-driven machine within manufacturing production as a whole – its ability to determine wages, prices and the overall nature of the system. What gave early industrialisation its revolutionary character, as historian Paul Mantoux noted, was not the machine *per se*, so much as what he called 'machinerism', namely the leading role assumed by mechanised production within the entire economic and social system[192]. Due to unprecedented increases in productivity reached with Arkwright and Crompton's devices, the so-called 'inanimate' machine – meaning its being animated by the non-human power of water – allowed a revolutionary increase in the power potential used by each worker, as well as a substantial increase in the amount of goods produced per unit cost. Such achievements were unthinkable in the proto-industrial world of biological energy, moved by human or animal muscle. The 'inanimate' machine became the very symbol of the new mode of production, in the dual sense of a Promethean instrument allowing the liberation of natural and social forces alike and of a new form of slavery and oppression: that experienced by working people being transformed from producers into machine operatives. The harnessing of natural forces met with general awe and social approval, while the mechanisation of labour was a much more contested phenomenon. What the process itself tended to hide, however, was the amount of human work still required by industrial production, both as muscle-power involved in attending inanimate machines and as skilled labour incorporated in a number of operations[193].

Indeed, even well into the industrial transformation, not all work was accomplished by inanimate power. To begin with, the factory system did not supplant proto-industry but worked alongside it. By mid-century, many wool industrialists in the Liri Valley owned both a mechanised mill in the valley and a non-mechanised workshop in Arpino (or some other rural village nearby), mostly running hand-looms. In other cases, industrialists assigned part of the work to domestic weavers, maintaining their previous role as 'putting-out' capitalists. The mountain-and-river system was far from dismantled by the Industrial Revolution: on the contrary, most manufacturers actually worked closely with one another, while economies of scale and scope were reached on a district level (rather than that of the individual large firm), benefiting from the respective comparative advantages of mountain and valley, such as the availability of mechanical power or skilled labour – and even of animal power. As late as 1865, eighteen workshops were still listed in Arpino: half of them reached roughly 278 hp by means of (real) horse power[194].

Though lasting throughout the industrial era, the political ecology of the mountain-and-river system had been substantially transformed. While in the time of proto-industry it was a mixed mode of production between feudal power and merchant capital, during industrialisation it became a unified mode where industrialists controlled both mechanised and non-mechanised labour, mainly weaving. The two forms of energy (the biological and the mechanical) were largely complementary, but their relationship was unequal: while human and animal-driven machines accomplished a substantial amount of labour, their power, size and productivity were largely inferior to those of the constant, uniform hydraulic machines. The relationship between manual and mechanical labour was continuously challenged by technical improvements and investments in new waterpower potential and machinery, through which industrialists aimed to increase productivity.

This energy history scenario, however, can only be fully appreciated in the context of the local ecology – of the existing social, gender and ecological relations of production and the agrarian environment around them[195]. In the early 1850s, the Liri Valley was a place where most people struggled to survive: while Isola Liri was experiencing a demographic explosion typical of industrialising centres – its population nearly doubled between 1812 and 1852 – Arpino and the other rural towns in the valley witnessed a slow increase, and some even lost population, in the same period. The overall result, according to historian Alain Dewerpe, was 'secular stagnation'[196]. Like proto-industry, industrialisation was a mere response to an un-favourable population–resources ratio in the marginal economy of a Mediterranean Apennine area, where agricultural land was scarce and productivity limited by lack of irrigation and low-quality soil[197].

Though substantially correct, this picture does not account for changes in social relationships that had occurred in the meantime. In the post-feudal age, patterns of land property in the Liri Valley were going through the definitive dismantling of common property [*demani*]: use rights and the commons were disappearing, while the agrarian economy was based on small-size tenant farming in sharecropping – a preferred form of land tenure throughout the Italian Apennines. Unfortunately, the relationship between this agrarian revolution and the industrial, in the Liri Valley, was very different from what was expected: while depriving local people of traditional access to subsistence means, agrarian enclosures had failed to produce a considerable increase in agricultural production. This failure should be ascribed to land-tenure systems as these emerged in the context of – and contributed to producing – the 'natural' conditions of production, that is, the agrarian environment.

According to Neapolitan writer Filippo Cirelli, who visited the Liri Valley in the early 1850s, available agricultural land was marked by a sharp contrast between the plain and the surrounding mountains. The mountain agrarian environment was a landscape of bare rocky wasteland, where only a few goats and sheep could find pasturage, while no crop could be grown. The lowlands, by contrast, (approximately 20,000 acres) featured olive groves (5,000 acres), 'cultivated plants' – mostly vines and fruit trees (10,000 acres), orchards (1,000 acres), and sparse woods of oak and chestnut trees (2,500 acres). Chequered with small farms and mixed with the variegated riverscape of the Liri and Fibreno, this landscape awakened 'surprise and enchantment' in the observer. Although intensely cultivated, however, the valley and low terraces could be much improved, the author remarked, by extending irrigation beyond the few plots along the riverbanks where it was practiced. Cirelli recommended that a system of canals winding from the riverbanks upstream from Isola – where the riverbed ran at a level higher than that of the countryside – be built by the Town and shared among irrigators by the payment of a fee. But such an innovation, he added, could only be accomplished by the authority of the Government, necessary 'to overcome the interests, rivalries, jealousies and gossips' of the landowners. Lacking irrigation, the overall production from intensive agriculture in the valley was not sufficient to meet the town's needs. Cereals (wheat, corn, oat and barley) were especially wanting, but even wine and the oil needed for wool manufacturing were to be 'imported' from outside[198].

The Liri Valley environment was thus dominated by labour-intensive agriculture (the 'land of trees', orchards and terraces) whose overall productivity, though, had not substantially changed since the feudal era. Despite largely passing into private hands, land had not been subjected to significant investment and labour was still the main instrument of agricultural improvement. Interestingly enough, half the bigger estates put on sale in 1868 were purchased by the industrialists, but this apparently did not imply any investment in agrarian 'improvement', such as irrigation and/or waterpower infrastructures.

Thirty years later, when the first agrarian survey of the kingdom of Italy was brought to the valley, local surveyor M. Mancini saw quite a different scenario. Irrigation was now practiced throughout the plain, but its effects were dubious and its management highly 'irrational'. A number of disparate ditches had been dug from the riverbanks of the Liri and minor tributaries in the plains downstream of

Isola, but proper distribution was lacking, so that lands were simply flooded for several hours a day. Since no levelling of the irrigated fields – a time-consuming, capital-intensive investment – had been carried out, much water was uselessly wasted and even damaged the land, which remained 'pale and barren' despite the great abundance of water, as if 'fleeced by the huge quantity of soil carried away' with the runoff. Conversely, many surrounding plots remained unirrigated for the same reason. An irrigation scheme, prepared by a local engineer and approved by the provincial government, had been set aside due to the opposition of landowners, who refused either to invest in the distribution dikes or to pay the annual government fee for the use of water. Once the project of such a public infrastructure had failed, irrigation rested upon the shoulders of poor share-croppers who lacked financial means and/or credit. Despite being potentially rich, agriculture in the plain of Sora thus merely compensated for the amount of work carried out by tenants and roughly half the population in the district had to work in the factories in order to survive. At the same time, the environmental effects of such irrigation methods had been disastrous both for the soil and for human health, as poor drainage led to a substantial increase in malaria and rheumatic fever in the local population[199].

Meanwhile, the industrial economy of the valley was showing its own cost in terms of human and environmental health. The ecological contradictions of industrial capitalism manifested themselves strikingly in the form of floods: while eight major inundations are recorded in the Liri Valley between 1825 and 1912, ordinary floods occurred annually in the rainy seasons, such that the two riverside factory towns came to live in a permanent state of near-disaster[200].

As in many other Apennine river-systems, floods had always been part of river ecology in the Liri basin. Yet, starting in the decade of French governance, their frequency and intensity increased substantially. After a disastrous inundation in 1833, occurring in the midst of a period of exceptional industrial growth, floods simply became a regular, even calculable, phenomenon. The reasons for this change in the flood regime of the Liri in the nineteenth century may be diverse: shifting rainfall patterns and melting glaciers due to the end of the so-called Little Ice Age may well have played an important role. Certainly, changes in political economy since the second half of the previous century, and especially agrarian enclosures upstream, perversely interacted with climate change to produce negative effects on the human environment.

Industrialisation, in turn, had produced substantial changes in the river-scape of the Liri Valley downstream. After running through the largely deforested landscape of the Apennines, coming to the walls of Sora the river found industrial capitalism, with its system of water enclosures, which had created an environment particularly favourable to the obstruction of the watercourse and the overflowing of its banks. Leaving on its left the mountain of Arpino, the Liri flowed into the complex weave of mills and villas that marked its landscape in the industrial age.

Improvement vs. Habitation

Its bed became filled with a dense patchwork of waterworks – weirs, stone walls and tree branches – designed to appropriate waterpower, literally enclosing the river into individual properties. Mills now dominated each piece of riverside land, where water and labour were put in the service of industry.

Within the walls of each mill, machines connected to water engines through a complex network of wheels and shafts were attended mostly by women and children working twelve to fourteen hours per day. Villas and mills crowded the river to the point of cutting it off from the urban landscape and public access. The production of the industrial riverscape was at the same time a process of spatial segregation and gentrification[201]. Our 1850s commentator on the Liri Valley landscape, Filippo Cirelli, was struck by the sharp contrast between the cultivated countryside and the urban–industrial environment. Isola had been forced to accommodate an explosive increase in its population over the past few decades, such that space, air and light were especially missed. The aspect of the town, Cirelli wrote, was 'miserable and of no gracefulness'; the roads were narrow and dirty; the number of dwellings was not proportionate to that of the people, who lived 'ill camped' in precarious and overcrowded slums; the streets lacked any drainage system and were covered with garbage. In addition, factory wages were particularly low and stagnating: in the 1840s, men earned 20–30 *grana*, women 10–12 (the equivalent of 1 kilo of white bread) and children 7–10 per day, working between twelve and sixteen hours a day. Twenty years later, despite the increase in productivity due to mechanisation, wages had not been raised, while the literacy level was still under ten per cent. Begging and thieving became common in times of industrial fluctuation, while brigandage returned in the political commotions of the early 1860s, when the valley was incorporated into the unified kingdom of Italy[202].

The demographic growth of Isola and Sora was linked to high birth rates rather than decreasing mortality. Paradoxically, in the 1870s the two factory towns registered mortality rates higher than those of Arpino: a noteworthy role in this state of affairs was played by the poor hygienic conditions of the factories' shop-floors, especially in the paper mills. About sixty per cent of the overall death toll by infectious diseases in Isola was caused by lung tuberculosis, which mainly affected the women operatives working in cloth selection. Poor environmental conditions outside the factory also contributed to put additional stress on the valley's residents. Between 1879 and 1880 a serious mortality crisis, caused by malaria, hit the area: the fever was probably related to the appreciable worsening of flooding in previous decades, but was undoubtedly aggravated by defective sewerage and poor nutrition. According to the 1886 enquiry on sanitary conditions in the kingdom of Italy, both Isola and Sora suffered from frequent malarial crises, typhoid fevers, tuberculosis and measles.

Despite the persisting trends of high morbidity and mortality rates, populations in the two factory towns increased by fifty per cent (Isola) and 35 per cent

(Sora) between 1871 and 1911, due to the unceasing attraction they exercised on the nearby rural villages. This phenomenon certainly had something to do with changes in land tenure patterns. The enclosing of common lands had taken another decisive step in the late 1860s, when about 3,300 more acres of formerly communal and ecclesiastical lands in the area of Sora were put on sale by the new Italian State[203]. Further expelled from the countryside, peasants did not have many alternatives to searching for a factory job – so Isola and Sora offered a local alternative to starvation. In 1865 the industrial workforce in Isola alone amounted to 2,400 people in 32 factories, eleven of which were *cartiere*, ten *lanifici* and seven machine-building and iron-making factories[204]. Roughly 3,000 individuals still worked in the wool factories of the Liri Valley in 1872[205] before a serious crisis hit the sector in the following years.

Together with the loss of a means of subsistence from a direct relationship to the land, industrialisation had increasingly subjected local people to the cycles and fluctuations of trade. Two major factories, namely Zino and Picani, closed down in the early 1860s, with a loss of 800 jobs, only a third of which had been recovered by the end of the decade. But the most serious crisis hit the *lanifici* a decade later: by 1878 half the woollen mills had closed with a loss of a thousand jobs, which were never recovered. The 1880s saw the growing importance of paper mills, partially compensating for the reduction of the wool mills: seven out of the twenty biggest factories operational in the Liri Valley were now *cartiere*, with a workforce of 2,247 (including 400 children), compared to only 700 still working in the *lanifici*.

With the fall of the Bourbon monarchy in 1860, the Liri Valley was subjected to the free-trade policies of the new Italian State and inevitably became connected to a larger international market system, where the decreasing cost of woollen cloth was related to the mechanisation of weaving and to the production of low-quality textiles (so-called 'renaissance wool'). Both innovations had stopped at the doors of the valley – but market integration due to the fall of protection barriers and steam-powered shipping had not stopped. Industrialists responded to the crisis in wool production by drastically reducing the adult male workforce in the factories, replacing them with children and by intensifying the work rates of hand-loom weavers – who were forced to provide more product for the same wage. This vicious circle of socio-ecological relations in the Liri Valley kept the wool industrialists from investing in innovation. It was technical stagnation, in fact – even more than the insufficiency of transportation, education and credit facilities – that led to the decline of the wool mills in the 1880s.

As for the paper mills, fundamental technical innovations had been introduced in the 1870s, when the production process was converted from rag to wood-pulp processing. Nevertheless, by the second decade of the 1900s, waterpower had become a major constraint in the crisis of the paper mills. Drought-induced energy shortages on the Liri became so frequent that steam-powered machines had to be introduced – and thus the increased cost of coal during World War I resulted in a new, long-lasting crisis; with no waterpower, the machines stopped and so did work. Out-migration became the main survival strategy for the people of the valley[206].

In fact, it was trans-oceanic migration that finally destroyed the age-old mountain-and-river system of interdependencies: more than 85,000 people fled the district of Sora between 1876 and 1911, the great majority after 1900[207]. The most striking fact is that migrants from the Liri Valley outnumbered those from all other districts of *Terra di Lavoro*, representing more than two thirds of the total. Added to the 'secular stagnation' of the population in the valley, this must be considered the most evident proof of the failure of both agrarian enclosures and industrialisation in raising the local standard of living by increasing per capita incomes.

The Machine in the River: A Pastoral Narrative

Understanding the ecology of industrialisation in the Liri Valley means taking into account not only the material reality, but also the cultural dimension of waterpower, i.e. the shaping of a new ecological consciousness, dominated by the industrial-capitalist vision of water.

Unfortunately, neither the (mostly illiterate) working class nor the educated local elite of the early to mid-nineteenth century have left written memoirs testifying to their visions of nature and industrialisation. Nevertheless, we have relevant sources that speak about the crucial place the Liri Valley took in the emerging new ecological consciousness of the industrial era. Travellers' memoirs, journals and descriptions of the valley, all written between 1819 and 1859, the central decades of industrialisation, are unique testimonies of how the nation (intended here to mean educated public opinion in the capital city) saw the Liri Valley and how the place itself contributed to the new political-economic culture of the post-feudal era.

Started by the Bourbon State as part of a project of economic reformism and of military concern, the road connecting Sora to Naples assumed an unexpected function in the following period: it transformed the Liri Valley into a place to visit. Isola with its waterfall, in particular, became the stage for a newly-invented narrative of the industrial riverscape. The first example of such a narrative is given by one Abbot Domenico Romanelli in 1819. While travelling to the Abbey of Montecassino, the clergyman decided to divert his journey toward the Liri Valley with the intention of visiting the ruins of Cicero's villa in Carnello. Once in Isola, he plainly recorded an 'active and industrious population, which makes an easy living from the manufacturing of wool and from agriculture'[208]. Soon, though, his attention was captured by 'the greatest and most impressive sight' represented by the two branches of the river mixing just at the bottom of the Ducal Palace, after falling for about twenty metres from the rock where the palace itself stood. He was so delighted by the waterfalls that he mentioned them in the very title of the book – *Viaggio da Napoli a Montecassino ed alla celebre cascata d'acqua nell'Isola di Sora* ['Travels from Naples to Montecassino and to the Famous Waterfalls of Isola di Sora'] – and asked a local painter to draw an illustration, or *veduta*, of the scenery, which he then included in his journal as 'one of the most graceful images of our kingdom'.

Initially, the Abbot's description of Isola resembles the usual iconography of the nineteenth century Italian countryside: the two 'marvellous' falls were said to create the beauty of the place, an enchanted isle with plenty of gardens and citrus orchards surrounded by a variegated landscape. He then headed towards the Fibreno River, searching for the ruins of Cicero's villa, as he was eager to contribute to academic discussion about the location of the ruins. But on his way to the antiquities he was forced to stop and contemplate modernity as manifested in the new form of the riverscape. What the Abbot saw while walking along the road between Isola and Sora, along the eastern bank of the Liri, passing through cultivated fields and orchards, was 'many new houses built for the wool mills, canals and plugs and many factories of useful and sought-after arts'. Once he reached the site where the Liri and the Fibreno merged, he was surprised by a 'novel sight': before mixing with the Liri, the Fibreno split into two branches, one of which split again, forming two small islands. On one of these stood the ancient monastery of San Domenico, the destination of his search for Cicero's ruins (some said it had been built over them). But it now held a completely different attraction for the Abbot:

> I stayed for a while contemplating other, new wool mills built there, and then headed
> to the other small island, called Carnello, where I first stayed to examine the paper
> and fulling mills erected there and then the ancient ruins[209].

By the end of the second decade of the century, a new industrial landscape had taken root within and alongside the agrarian and the literary. The mills that the Abbot saw in Carnello belonged to the *Cartiera* of Santa Maria delle Forme. A few years before, the original plant had been enlarged by adding some structures on the opposite bank, on the islet in the middle of the river. The woollen mills that the abbot visited on the isle of San Domenico, once part of a monastery, were now just some of the many factories in the valley, established in 1816, inside the former convent of San Francesco near Isola Liri.

The importance of the recent shift to economic liberty for the economy of the area and potentially of the nation was theorised and celebrated by the highest authorities in the kingdom well after the end of French rule and was perceived as a progressive change by other contemporary observers. One eloquent testimony to this new consciousness is a 'pictorial' description of Isola Liri published in 1829. Written as a comment upon an illustration of the town dominated by the waterfalls, the text started with a triumphal paean to the wealth that rivers offer 'civilised nations which, running those waters to their own benefit, benefit from them [...] as power sources for one thousand kinds of machines'. Progress, the authors said, is the

The Machine in the River: A Pastoral Narrative

Figure 5. R. Carelli, 'Cascata dal Fibreno'. In Domenico Cuciniello and Lorenzo Bianchi eds, *Viaggio pittorico nel Regno delle due Sicilie dedicato a Sua Maestà il re Francesco primo*, (Naples: SEM 1971 [1829]). Courtesy of Ministero per i Beni e le Attività Culturali, Biblioteca Nazionale di Napoli, Sezione Lucchesi Palli. All rights reserved.

very ability of drawing from river-flows a great 'increase in industry and wealth and prosperity'. An emerging mystique of waterpower, grounded in the peculiar Italian landscape, was developed by the authors when depicting the Liri and Fibreno rivers as 'spreading with their humours power and energy and prosperity', so much so that it had become impossible to say whether their fame was linked to their history and 'the natural beauty of which they form such a great part' or to 'the maximum utility they bring to the people living along their banks'.

The authors go on to describe the course of the Liri down to the isle of Carnello, where, 'running among gentle falls, it then narrows into a little pond, from where, splitting into several canals, it gives shape to very graceful small isles, joined by country bridges.' The image is one of pastoral simplicity and idyllic beauty:

> All this place looks like a very precious garden, made from art less than from nature; and its main ornament is a long, twisty, delightful boulevard, offering one of those promenades that are now called *romantic*. Thousands of variegated and pretty sights are enjoyed from it; but none equals that of the so-called *Cascatelle* [Little Falls]: before passing through the canal of Le Forme, as if announcing its big fall, here the Liri playfully breaks among declivous rocks, lying with peculiar irregularity in the shape of five stairs, among trees and leaved bushes. The water, rebounding, rumbling, foaming and spreading in white flakes, finally merges in a short, regular fall.

In this pastoral narrative of the river's gift, the authors created an icon of the national path to industrialisation, one in which art and nature, history and progress are wholly merged and cannot be distinguished. They concluded:

> And, as if beauty should never, in this happy district, be separate from the useful, those falling waters have already moved the wheels and cylinders of the wool factory that currently honours the ancient palace of the Dukes of Sora, who sold it, along with all their possessions, to the Government; and the Government gave to the national industry[210].

Here at last is a reference to those power changes that gave birth to the process of industrialisation in the valley. What the authors call the 'national industry' was in fact the *lanificio* of Lambert, one of the most powerful industrialists in the valley. The narrative of the industrial landscape blends invisibly the natural and social forces acting within the local environment. It contributes, in its own way, to legitimising the current assets of power by locating them within a natural order that has been restored by politics. Rescued from feudal possession, and after the government in its wisdom has assigned them to the best possible use, i.e. private industry, the Liri and Fibreno can show their maximum beauty *and* utility, contributing to the glory of the nation.

In this sense, the industrial riverscape narrative is also one of political economy, since it appropriates the language of that discipline and translates it into a broader social discourse of economic and environmental change. In 1837, the *Poliorama Pittoresco* (a popular journal of culture in the capital) published a description of Isola Liri that sounded like a manifesto of economic liberalism[211]. Inserted in a section devoted to travel reports from the provinces of the kingdom, the article starts with the usual ecstatic account of the waterfall and the landscape, depicting a 'sunny valley of pleasant hills'. Yet this is just a prelude to the 'hollow and monotonous sound, which [is] the roar of the falling water and the incessant beating of fulling mills, processing paper, cloths and other works in this district'. Perhaps for the first time in the century, the beauty of the waterfall is linked not so much to its rural–natural character as to its industrial–artificial one. Despite depicting it as a 'water hell', the author finds the fall to be the natural backdrop for a 'very commendable town with its extremely industrious and active population'.

The article invokes quite contradictory images of memory and change, romantic ecstasy and industrialist emphasis. In the workings of the paper and wool mills, the author sees 'the principles of political economy, *this sublime science* of the eighteenth century, accomplishing their sacred purpose'. Testimony to this new economy of nature is the 'ancient tower' (the Ducal Palace), 'no longer a stage for oppression, but a sacred, honourable workshop, where industry and the mechanical arts concur in paper and cloth manufacturing'. He then wanders through the 'labyrinth of stairs and rooms, under the dark subterranean vaults where the noisy splash of the water and the monotonous fulling of the machines echo' and then, to enjoy the contrast, heads to the foot of the second fall, where the water, 'its fury and foam lost, flows through a bed of solitary and silent sand'. There, at the end of this verbal tour of mechanised industry, Isola Liri appears to him as 'one of the most romantic lands of our kingdom'.

The language of political economy and of the romantic sublime intertwine in this text to forge a new vision of the landscape, narrating a story of progress: from feudalism to industry, from the quiet idyllic rural landscape of the past to the industrious town where capitalism can finally be celebrated within the walls of the factory. And yet, no real contradiction can be seen in this vision beyond the aesthetic contrasts so attractive to a romantic sensibility. The beauty of nature lies precisely *in* its incorporation within the factory system, not in an idealised past. The most striking feature of this account is its ability to 'naturalise' both social and environmental changes that have occurred in the place by dissimulating the oppressive character of industrialisation towards both nature and humans. The palace that was once home to feudal power is now a stage for a more modern form of oppression, that of industrial labour discipline, forcing women, men and children to follow the rhythm of that same monotonous beating of machines under water that our author finds so exciting. Furthermore, water itself has become subject to a process of increasing domination, the Boncompagni Palace being only one site among dozens of mechanised factories along a short stretch of the river basin benefiting from the availability of waterpower.

This translation of political economy through the quiet and reassuring language of the pastoral landscape narrative is demonstrated in a number of other 'secondary' literary texts produced in the same period. In 1845, on the occasion of a conference of Italian scientists held in Naples, the economist Matteo De Augustinis described Isola Liri as 'the Manchester of the Two Sicilies'. His account of textile and paper manufacturing in the Valley is a celebration of the industrial landscape in its most artificial version and a powerful exercise in environmental mystification. To him the valley is in fact 'a huge and whole manufacture; so many are the buildings and workshops and many the mechanised factories'.

> Noise and the sprinkling of water [...], the creaking of machines and wheels, the sight of the exploited water, which has *become imbued with thousands of colours by the variety of dyes*; encountering endless wool and cloth, rags, piled-up paper; the encumbrance of carts and barrows in all the streets, in all directions; everything you see shows that you are in *the valley of labour and industry, that once was of leisure, ease and study.* (My emphases.)

Here another step was taken towards creating a positive, progressive narrative of environmental and social transformation, one that shrouds potential ills in pastoral rhetoric. In two extremely powerful sentences the author manages to sublimate the polluting effects of the paper mills as a sign of the increased ability to dominate nature, impressing it with the colours of the new production system;

and to sanction the definitive shut-down of a past leisure time and its replacement with work.

In the course of this translation of the realities of political economy into a new narrative of the prosperous industrial landscape, water itself was also subjected to an increasing process of de-materialisation, one that progressively transformed it into an abstract, mechanistic and atomised commodity. The Liri and Fibreno rivers were now compared to gold mines in the New World, being 'in the scientific terms of economics' more valuable than the latter[212]. The German historian Ferdinand Gregorovius, travelling through Italy in the late 1850s, saw a golden opportunity in harnessing water: the paper manufacturer Carlo Lefebvre, he noted, arriving in town without means, had, 'by carving pure gold from the force of water, transformed the Liri riverbanks into an Eldorado', gained the title of Count and left his son Ernesto with 'factories and millions'. The language of the industrial sublime is employed again and again to describe the 'magnificent and gracious buildings' of the two Lefebvre paper-mills and the Count's garden villa along the Liri riverbank, improved with canals and resembling 'a little Tivoli and a paradise of nymphs'[213]. The Liri, like the garden that watered Eden, is a celebration of the pastoral paradise of water[214].

Epitomising the industrial narrative in the Liri Valley, Gregorovius depicts a new logic of the pastoral sublime. The river basin is, to his eyes, an enchanted site and the Liri 'developing its green water amongst high poplars' is 'quiet and sleeping'. Its banks are 'enchanting, melodic, sunny and dreamful'; by contrast, the town of Sora, 'now so modern', is described as 'having a good road, an industrial life, animated traffic'[215]. On his way along the Liri toward Isola, the author discovers the presence of an industrial elite, indicated by the 'delightful villas and their industrial workshops looking out from among the trees'[216]. Although the traveller is mostly attracted by the beauty of the aquatic landscape, the sound of the waterfalls, the sight of the innumerable canals running swiftly into the river and the 'marvellous vegetation so typical of southern countries', he nevertheless notes that the plentiful water which runs around Isola gives power to many manufacturers, forming 'a robust colony of factory workers' and producing a beneficial effect throughout the district. 'Blessed by Nature', the author concludes, 'Isola will always be the leading industrial town of the area'.

The emergence of this particular narrative should be read in the context of waterpower technology. Unlike the American locomotive of Leo Marx's narrative, the Italian machine in the river did not look like a counterforce, resembling the violence of industry 'as a memento of reality'[217], but like a longstanding part of it, perfectly inserted into the landscape as a sign of how humans can live in harmony with nature while using it; in other words, how capitalism and nature could be harmonised in the pastoral-industrial landscape. Indeed, Italian industrialisation took place in riverscapes that did *not* resemble satanic places and the machinery belonged to a mill that was often a remnant of the past (a feudal palace, a former monastery), symbolising how this harmony between work and nature embraced history as well.

What these industrial landscape narratives have in common is a new ideological horizon: the need to build an acceptable social landscape in which history is not cancelled, nature is not mortified, but put to its proper use and industry does not represent a 'satanic' agency, but a virtuous path toward the public good. They share a pastoral ideal of industrialisation that harmonises beautifully with the Italian physical and social landscape: a landscape in which factories and agriculture are included in the same view of one river valley, that can also encompass ruins, poetry and literary memories, and the observer can enjoy a full contemplation of history as the passing of time from one form of civilisation to the next; a landscape where there is no place for the exploitation of either labour or nature and where even the alienation of people from nature seems not yet to have occurred; a place, finally, where the appropriation of rivers by mill owners is a wise achievement of modernity, conducive to prosperity for all[218].

***

Viewed through the lens of socio-ecological relations, a fresh perspective on the 'Industrial Revolution' on the European periphery can be gained. First, proto-industry was a mountain-and-river system, shaped by socio-ecological relations of power: feudal control over water and land, capital control over labour. Second, industrialisation came to incorporate waterpower within the factory, producing new forms of social domination over both labour and nature.

Nevertheless, rather than the classic imagery of factory towns, Isola and Sora in the nineteenth century evoke that of ambiguous socio-ecological realities, where the paternalistic values of a peasant community long dominated cultural and economic behaviour. This ambiguity, moreover, was not a transient phase, but the very essence of the local industrial landscape. Not only would the mountain-and-river system continue to exist for a long time, the handlooms of Arpino being connected to the mechanised spools of Isola, but the riverscape itself testified to the incompleteness and hybridity of the industrial transformation. Having remained substantially unaltered by dams, diversions, navigable canals and the like, the course of the river was still the same as during the previous mode of production. Water-wheels and mills were now far more numerous and so many waterworks obstructed the flow, while more mountain soil was carried by the river and deposited along its tortuous course downstream. Nevertheless, this was perceived to be the same river that had crossed the valley 'since time immemorial', on whose banks it was still possible to search for the ruins of a Roman villa. Although industrial capitalism dominated the riverscape with its presence and controlled the socio-ecological relations of production, it did *not* dominate nature in its entirety. It did not control the flow, direction and quantity of water allowed to run through the valley; this only happened much later, with the advent of the hydroelectric plant, entailing a completely different political ecology. Industrialists did not even have enough influence to stop the process of deforestation and tillage upstream that caused the mills so many problems.

The hybridity of the Industrial Revolution in the Liri Valley is also represented by another socio-natural feature: waterpower being a land-based energy source, the capital gain that it generated, once incorporated within the factory system, could be assimilated to a form of land-rent. As the owners of waterpower, mill-owners exploited the free labour of nature without reinvesting any of their profits in energy conservation, or in preserving the local hydro-ecosystem. In fact, as the next part of the book will show, both the social and the environmental costs of industrialisation in the Liri Valley can be related to the lost opportunity of managing waterpower as common property, using it to the benefit of the community and with greater respect for hydrological equilibria.

The weakness of the Industrial Revolution in the Liri Valley was perfectly reflected in the unique narrative that emerged in travellers' accounts: this was not so much a narrative of the industrial sublime and mastering of nature, as one of pastorality. Stemming from the political economy of anti-feudalism and incorporating the Enlightenment project, the pastoral history of industrialisation in the Liri Valley was in fact one of liberation, not of domination. Nature had been recovered from the 'fetters' of feudalism, such that an Edenic relationship with it could be finally recreated. The pastoral narrative was a crucial element in the making of the ecological consciousness of Italian capitalism in the industrial era, for it persisted within waterpower technology and mountain-and-river systems for the entire nineteenth century and beyond. Even the peculiar nature of Italian environmentalism, compared to that of other industrial nations, can be traced back to this pastoral origin story of industrial Italy.

Indeed, the Liri Valley narrative of industrialisation is characterised by its complete lack of dissonance, its perfect contentment with the landscape it describes. This was obviously a re-elaboration of the theme of pastorality, which originally implied a sense of nostalgia and the search for refuge in nature[219]. Here, in contrast, the emphasis is on the present and future promise of the new landscape, while the past is the place for social domination and discord. The movement of the waterwheels in the river represents the essence of this harmony: there was no steam-powered vessel coming up the river to interrupt the idyll with its disturbing whistle, nor smoke pouring out of chimneys and polluting the clean air of the valley. Despite violently interrupting the Edenic harmony of the landscape, floods played no part in the industrialisation narrative. There are many possible reasons for this, not the least of which is the fact that those who produced the pastoral narrative did not live in the place they mythologised; as travellers, they were part of the landscape only for the duration of a transient gaze[220]. Nevertheless, floods and the other social costs of industrialisation were part of the everyday life of local people and marked the way in which several generations experienced the transformation of the mountain-and-river system in the post-feudal era.

In the space of roughly a century since enclosures and mechanisation first appeared in the valley, a revolution in socio-ecological relations had been completed. Contrary to expectations, this revolution had not improved either nature or social life and had resulted in massive out-migration. Some economic historians would explain this paradox in terms of inefficiencies in the overall productivity of energy sources (both biological and mechanical) and in the population–resources ratio[221]. Others would point to the inconsistencies of institutional and entrepreneurial behaviour[222]. If we view industrialisation as the product of interaction between social and ecological forces, both schools seem to have relevant things to say about the Liri Valley case. The idea informing this book, however, is that something is missing from previous explanations of industrialisation and that this has to do with the political ecology of the mountain-and-river system: that is, the distribution of property and environmental costs.

In the Liri Valley, while the productivity of industrial labour increased significantly with the shift from proto-industry to the factory, land productivity did not increase in the same period. Little of the profit that entrepreneurs accumulated in the manufacturing sector by exploiting both water and labour power was reinvested in technical innovation and/or in what economists call positive 'externalities', such as education, infrastructure, transportation means, credit facilities and the like. In this respect, Isola at the end of the nineteenth century did not look very different from half a century before, when the factory system had taken shape in the local riverscape, reaching its spatial limit. In the first half of the century, individual appropriation of water had allowed a significant primitive accumulation to the benefit of a new social group, the manufacturers, who had materially taken the place of the feudatory and the Church. At the same time, land was in the process of being appropriated too: land productivity, though, did not seem to increase significantly in the aftermath of agrarian enclosures. Irrigation had remained limited and ill-practiced, while both uphill deforestation and industrial development downriver had produced a deterioration of topsoil quantity and quality and a significant increase in flood risk/vulnerability. Furthermore, as in different European contexts, the industrial bourgeoisie tended to reinvest its profits in agrarian estates in order to consolidate its social prestige, which remained entwined with landed property throughout the nineteenth century and beyond. As landowners, industrial entrepreneurs manifested even less innovative behaviour than they did as mill-owners, such that not even these newly enclosed strips of the Liri Valley experienced substantial increases in productivity. The overall socio-ecological nature of the industrial transformation of the valley was that of parasitic exploitation on the part of a narrow minority over both nature and labour. While local people were dispossessed of direct access to land and water, also losing control over their own labour in the process, what the appropriators gave back was meagre and stagnating salaries, partly in food-crops, which merely compensated for the lost income of domestic production and the end of the moral economy. Throughout the nineteenth century, 'survival' remained the key word for generations of peasants and factory workers in the valley. No 'economic development' can be seen in the area, only the shift of energy rents – both biological and mechanical – from feudal to capitalistic control[223]. Such a shift, however, did have its cost.

The social costs of industrialisation in the Liri Valley do not substantially differ from those experienced elsewhere during the 'great transformation'. The factory discipline replaced domestic labour. The direct relationship of humans with nature as the means of subsistence and production was alienated, as was the relationship of workers with their own bodily energy and mental skills. Environmental and health vulnerability significantly increased, as did inequality in the social distribution of risk[224]. Children were overexploited. There was discrimination against female labour.

Nevertheless, the social costs in the Liri Valley did have their historical peculiarity: they exceeded benefits, in the long run as well as the short term. Social control, the patriarchal order, illiteracy and the lack of opportunities so typical of the *ancien regime* retained an unaltered grip over the local community. At the outbreak of World War One, after roughly one hundred years of industrialisation, the Liri Valley could still be regarded as a peasant society, rather than an urban industrial world[225]. At that point, however, the poor would cease paying the bill of industrialisation by using the only means at their disposal: leaving the valley.

Thus at the end of the century neither land nor water enclosures had brought about the improvement in nature that enlightened philosophers and State agencies had theorised. An overall imaginary cost–benefit analysis of the industrial transformation in the Liri Valley is negative. It is important, however, to recognise that this *is* a case of industrialisation, similar in many respects to that occurring in the same period elsewhere. This chapter sought to show how extensive and radical the changes were in the local landscape, as well as in modes and relations of production. Unlike many other Mediterranean mountain valleys, it was neither the lack of agrarian capitalism nor of the factory system that prevented the Liri Valley economy from becoming rich.

How then do we make sense of this failure of political economy to keep its promises about the beneficial effects of enclosing the world? The following chapter will seek the answer at the very core of the new relationship with nature that industrial capitalists established in the valley, centred on their particular interpretation of the discourse and practice of private property.

Part II

The Economy of Water

'The history of the commons in southern Italy is that of an uninterrupted series of usurpations'

Parliamentary enquiry on the conditions of peasants in southern Italy and Sicily, 1911[226]

One Hundred Years of Enclosures

With the fall of Napoleon at Waterloo in 1815 and the defeat of Gioacchino Murat at Tolentino, the 'French decade' had come to an end and the Kingdom of Naples had been returned to the Bourbon King Ferdinand I, reincorporating Sicily and taking the name of Kingdom of the Two Sicilies[227]. Not much changed, however, in the political organisation of the State, which simply validated the laws introduced by the French rulers, including the abolition of feudality. As time went by, it became clear that the country was frozen in a situation of potentially explosive social conflict, whose core was the land question. In fact, much of the historiography of southern Italy (the so-called *Mezzogiorno*) has identified in the never solved '*questione demaniale*' – the problem of partitioning the commons [*demani*] into small individual properties to be assigned to the rural poor – the main source of political/economic backwardness and social unrest for a very long time, stretching from the abolition of feudality to the post-war period[228].

The question of common lands is probably the most significant example of how nature and political economy became connected in nineteenth century southern Italy, through the dialectic between discourses and practices of appropriation. Indeed, in its few years of activity, the Feudal Commission instituted by Joseph Napoleon with the purpose of dividing the *demani* between ex-feudatories and communities – leaving to the latter the possibility of partitioning the commons among the peasants – had not been able to terminate the infinite array of litigation converging around the common lands of southern Italy[229]. The uncertainty of rights, whether of property or use, long continued to be a peculiar feature of social life in the countryside and opened up infinite possibilities for social tension, litigation and political patronage[230]. For a number of reasons, the project of 'land individualism' originally entailed by the Enlightenment project was not fully implemented. A great quantity of common land throughout southern Italy remained undivided in the form of communal *demani*, belonging – or entrusted – to the *Comune* (i.e. the

Town Corporation). Part was left to the local poor for their use rights, regulated by the *Comune* according to a variety of local customs; part was rented out to tenants in order to increase cash revenues. In a sense, the history of the rural South in the nineteenth century can be seen as a long story of conflicts over the use and posses- sion of such *demani*. Crucial to the moral economy and survival of the peasants, the *demani* were an obscure object of desire for the emerging rural bourgeoisie, always struggling to secure exclusive access to local resources as a means to social and po- litical power. Ideologically, the permanence of the commons and of undivided use rights was considered a sign of backwardness and a waste of resources. In practice, it helped local elites to illegally enclose as much as they could of the *demani*, only raising the sense of injustice and usurpation in the powerless and illiterate *cafoni*. In addition, through the possession of land, the rural bourgeoisie had gained ac- cess to local government. According to many scholars, the failed partitioning and the private usurpation of the commons were due to the ability of the new agrarian elite to control local politics, thereby preventing the formation of a class of small independent farmers.

Even when actually implemented, however, the division of the *demani* had quite different results from those expected. Its main effect was to convert more and more public land (and water) into private property, which was sold to wealthy rural groups, with substantial losses for the poor in terms of access to resources. In fact, the land-ownership ratio in southern Italy remained persistently low throughout the nineteenth century and up to the mid-twentieth, when an 'agrarian reform' was finally implemented[231].

A living proof of how harsh social oppression had become in the provinces was the diffusion of the brigands. From occasional episodes linked to war and revolution – as mentioned in Chapter 1 – rural brigandage and its urban variants had become a permanent feature of the southern Italian landscape, reaching higher organisational levels and massive dimensions in some areas[232]. Albeit a complex and contradictory phenomenon, brigandage can be considered an expression of violent protest against processes of change perceived as unjust and disruptive of social cohesion: in the rural countryside, such a perception of injustice was almost invariably linked to the results of de-feudalisation and the related partitioning of common lands. Loyal to the royal family and firmly attached to the Church, which often gave them shelter and protection, southern Italian brigands did not act in the name of a specific political programme. By exercising a mixture of terror and revenge, brigands often obtained support from the local population and were sometimes the object of reverence and cult. On the other hand, they were themselves used by the same rural middle classes that were 'civilising' the provinces, who were no strangers to using brigands to settle family *faide* [feuds], or to secure their exclu- sive control over local communities, natural resources and politics. Brigandage, in other words, testified to the very contradictions of modernisation as experienced in southern Italy, starting with French rule. After the unification of the country, these contradictions became a real war, fought between the Italian army and the brigands – but also involving the local population, without much distinction, in the course of the years 1860–65.

In May 1860, Giuseppe Garibaldi's 'enterprise of the *Mille*', an expedition of a thousand men, sailed from Genoa to Sicily and from there marched up through the Two Sicilies, 'liberating' the area from Bourbon rule. The armed expression of a romantic national ideal – the Italian *Risorgimento* – the *Mille* found enthusiastic support from southern Italian peasants, to whom they promised the assignation of land plots, and in a few weeks the State collapsed. The rapidity of the fall of what was the largest nation in pre-unification Italy can be explained by both the willingness of southern Italian elites to coalesce within the new political system[233] and the existence of harsh social conflicts, which Garibaldi's venture brought to an open deflagration. During the Restoration, social conflict related to the question of *demani* had occasionally exploded into violence, especially during periods of political disorder, such as the 1820–21 and the 1848 'revolutions'. However, widespread armed resistance took root across the countryside and forests of southern Italy soon after political unification in 1860, when it became clear that no change in land tenure patterns would be permitted by the new rulers. While the unification of the nation had been mostly a bloodless venture, a real war – lasting five years and leaving thousands dead – was that which started immediately afterwards between the Italian army and southern brigands[234].

As happened between 1796 and 1806, the district of Sora was again thrown into the midst of violence. The brigand Luigi Alonzi, called Chiavone, a descendant of a certain 'lieutenant' of the brigand Mammone (see Chapter 1), controlled the area between Sora and Veroli, across the border with the Papal State. Interestingly enough, Alonzi used to be a forest keeper in the Water and Forest Administration. According to a French reporter, who was able to reach him in his mountain retreat and publish an interview with him in *L'Illustration*, Chiavone had gained local respect (and a certain fortune) by allowing poachers into the forests of Sora[235]. After 1860, he had assembled an armed band and taken to the mountains, where he acted in the name of the 'legitimate' Bourbon King Francis II. This shows how, contrary to common perceptions, brigands (and their admirers) shared the same romantic culture that fed Italian nationalism – but on the flip side of history. Chiavone, who had great admiration for Garibaldi, became himself a symbol of anti-Piedmontese revolt and local identity and his cult is still lively today[236]. The Piedmontese, however, did not allow brigandage to be seen as a nationalist struggle; indeed, contesting the very identity of the enemy was part of the war strategy. Predictably, one consequence of the war on brigandage was its contribution to the making of a newly invented Italian identity, one in which both nature and the people of the south played the role of wild 'others' to tame and then civilise[237].

Once the Piedmontese military had withdrawn from a pacified South in 1866, a new State infrastructure, with its civil servants and politicians, replaced them. Some of these were former Neapolitan liberals, who had escaped the ferocious post-1848 Bourbon repression and censorship by fleeing to northern Italy or Europe. While exiled, they had elaborated a strongly negative vision of southern Italian society as one anchored to a pre-modern and illiberal culture[238]. The 'southern question' of the united Italy was then officially born when Pasquale Villari (one of the Neapolitan *patrioti* exiled to Florence) published his *Lettere meridionali* ['Letters from the South'], a book considered the very manifesto of so-called '*meridionalismo*' (pro-South politics). A group of writers and social scientists representing moderate, right-wing liberalism, as opposed to democrats and radicals, took pains to expose the 'southern question' to national attention through the pages of the Tuscan journal *La rassegna settimanale*. In the end, the group offered an image of southern Italy as an archaic social reality, where feudalism had been reincarnated in the *latifondo* (the large agrarian estate) and both people and the land were still awaiting a real liberation – which would become a permanent mission for the new State.

An object of repeated attention, for their embodiment of the very essence of archaism, became common lands and mountain communities, especially those situated in the Apennines of central and southern Italy. Starting in the 1870s, a new series of surveys, pamphlets and statistics were produced, the most important of which was the general agrarian survey known as *Inchiesta Jacini* (see Chapter 3). Among the relevant details on the different 'conditions of the agricultural class', the survey reflected an unexpected – and highly disliked – socio-ecological feature of the Italian countryside: the widespread persistence of undivided *demani*[239]. As a consequence, the 'question of the commons' became the focus of a huge juridical debate and the subject of relevant legislative activity between 1884 and 1894[240].

In the twilight years of the nineteenth century the south, officially renamed the *Mezzogiorno*[241], had definitely become a national problem, at the heart of which lay the peasants and the question of the commons. By emphasising backwardness, both parliamentary enquiries and political pamphlets seriously underplayed the major social change that had occurred in the recent history of southern Italy. Indeed, for roughly a century before political unification of the country, capitalism and political economy had made their decisive entry into both the city and the countryside, through massive privatisation of land and water and the progressive reduction of the moral economy. Long before entering the new national market, the south had been largely integrated into the commercial and bourgeois revolutions[242]. A look at the propertied classes and the socio-ecological context of their existence is thus necessary to have a better grasp of the 'southern question'.

A rough dividing line could be drawn between the two basic types of propertied classes that were emerging in mid-nineteenth century southern Italy: a line based on access to water. On one side lay the *latifondo*, the arid great estate largely devoted to extensive cereals and pasture. The repeated selling of public land, especially that confiscated from suppressed monasteries from the late eighteenth century onwards, together with usurpations and land speculation, had put a substantial amount of property into the hands of a new agrarian middle class, the *latifondisti*. A new socio-ecological complex, the *latifondo* spread across inland areas, owing its long-lasting success to its ability to meet the multiple needs of both local society and ecology. A hybrid figure between the great landowner and the feudalist, the *latifondista* managed his or her property with mixed commercial and moral economy objectives, thus obtaining social cohesion and the total subordination of the peasants[243].

On the other side of the water line lay irrigated farmland, largely devoted to commercial crops. Another major novelty in the socio-ecology of nineteenth century southern Italy was the boom in citrus fruits, olives and grapes, almonds and hazelnuts, mulberry and sugar cane, which substantially changed the agrarian landscape of the south, increasing the presence of fruit trees and shrubs (the so called *coltura promiscua*, or polyculture). In the river valleys and floodplains of nineteenth century southern Italy – areas such as Palermo's 'golden valley', the plains of Cirò and Reggio in Calabria, the Sarno floodplain near Naples, Terra di Bari and Terra d'Otranto in Apulia – new commercial bourgeoisies began to emerge as socio-ecological formations, based on the appropriation of both land and water[244]. The new plantations also contributed to changing the country's economy into that of a semi-periphery of the Atlantic world, strongly dependent on the demand for specialised crops in North European and the North American markets. While most of the commerce in grain and oil was controlled by a small group of speculators, partly foreign, who resided in the capital city and from there controlled the country's agrarian production[245], the growth of export agriculture brought about the rise of a new merchant bourgeoisie, based in coastal cities and in medium-sized agro-towns inland[246].

Following the Age of Revolution, one known as the Restoration (1815–1860) was a period of noteworthy changes in organised life throughout the Two Sicilies: urban life, in particular, experienced important growth and many cities were completely remodelled during this period, gaining their modern appearance. New streets and residential areas, fountains and cemeteries, theatres and libraries, schools and colleges, public buildings for government functions and a general façade of bourgeois décor were added to most cities and towns throughout the country. At the same time, 'public opinion' began to take shape through the foundation of new journals and a substantial increase in sociability in its various forms, from economic and cultural associations to cafes, spas and other public places for social intercourse[247].

More often than not, however, modern conveniences had made only a scant and formal appearance in the rural provinces. The Bourbon economic policy was basically a hands-off strategy with little taxation and even less spending: as a consequence, economic freedom was the rule, but infrastructure – both commercial and social – was particularly wanting. Extremely poor budgets and chronic State insolvency were behind the lack of investment in public works such as local roads, canals and irrigation schemes, schools, town halls, hospitals, etc. Schooling, in particular, was a major victim of this lack of care for the public: an 1831 decree, for example, expressly authorised insolvent town councils to reduce primary education, starting with girls' classes[248]. Health infrastructures were also dramatically lacking, as became evident during the several cholera epidemics that hit the country during the century. Furthermore, the Bourbon low spending policy had important repercussions in socio-environmental terms, especially as regards the question of *bonifica* – Land Drainage and Improvement.

Water and the State in the Mezzogiorno

The 'disorder of water', as this evolved after the end of the French period, is probably the most relevant terrain in which to observe the redefinition of 'public' and 'private' spheres, along with the marginalisation of the 'common', characterising this period of southern Italian history. The history of Water and Forest and Land Drainage and Improvement bureaus during the Restoration shows what a deep and inextricable relationship was being set up between nature and political economy, in both discursive and material ways[249].

Chapters 1 and 2 showed how floods and the 'disorder of water' had become a recurrent concern to Neapolitan reformers at the end of the eighteenth century, when the capital city's intelligentsia acquired a renewed interest in the 'natural resources' of the kingdom. During the following period of revolution and imperial dominion, a completely new government infrastructure was created over southern Italy's society and nature. A 'Water and Forest' Administration was set up in 1811 as part of the new State bureaucracy, for the first time organically connecting the provinces with the capital city through lines of authority and communication. A web of peripheral articulation was created between the centre (the Minister of the Interior, responding to the Crown) and the periphery, through the provincial Prefects [*Intendenti*] who in turn liaised with the *Sottintendenti* (at the district level) and the Town Mayors; at each level, collegial organs were also formed: State, Province, District and Town Councils. A modern State had emerged from the process, surviving the Restoration phase and becoming an undisputed form of territorial government.

This highly centralised communication structure allowed an unprecedented flow of information concerning every single aspect of social and economic life in the country; it left in its wake a cornucopia of written documents that can still be materially experienced by visiting any of southern Italy's State archives. Documents concerning water use may be found in the archive section named *Economia di acque* [Water Economy], whose folders contain varied documentation regarding irrigation – including the maceration of hemp and flax and rice cultivation – waterpower and fisheries. Most of the information we have on waterpower use in the Liri Valley comes from this archive section. But the Water Economy files do not tell the whole story. Documents concerning floods and malaria were filed under the rubric of 'Water and Forest', which was part of a different archive section, called *Ponti e strade* [Bridges and Roads]. Long-forgotten information is hidden within this archive section, referring to the recurrence of floods in the nineteenth century; although largely inconsistent with today's measurement criteria, the very recording of the events holds intrinsic value, for it helps us recognise one important aspect of past reality that other sources might completely overlook.

'Water and Forest' files carried with them a reverse version of the 'Economy of Water', as this had been reconfigured with enclosures and improvements from the early nineteenth century. The files were mostly brought together by local civil servants, who kept complaining how much denser the forests were in earlier times, how much more frequent landslides and floods had become, how much scarcer wood was, etc.[250]. The widespread wisdom regarding water–forest connections inextricably linked the problem of the 'disorder of water' to the question of *bonifica*: the latter became the main terrain for 'public vs. private' political discourse in the Two Sicilies[251].

The man who, for almost thirty years, came to embody the politics of water-and-forest in the Two Sicilies was called Carlo Afan De Rivera (1779–1852) and was a hydraulic engineer. His *Memoria intorno alle devastazioni prodotte dalle acque a cagion de' diboscamenti* ['Report on the Devastations Produced by Water in Consequence of Deforestation'], published in 1825, embraced Monticelli's vision of the 'disorder of water', based on deforestation as a mono-causal explanation[252]. Published soon after the start of De Rivera's appointment as the Chief of the Corps of Bridges and Roads, the report reads like a statement of beliefs and intents, rather than the result of long-term observations. The author maintained that 'beneficent Nature' had provided the country with forests so that rainfall did not damage the land and rivers did not become torrents, but Man had reversed this order by denuding the slopes. De Rivera also raised the unquestioned narrative of the once-navigable rivers of southern Italy (created by the previous generation of writers) as proof of environmental decline due to long-term deforestation. Man, the disturbing agent of nature's order, was the main protagonist of De Rivera's report; nevertheless, times had changed since Monticelli's 1809 'On the Economy of Water' and so had politics. In De Rivera's version, the 'disorder of water' was caused not by feudalism and communalism (these having largely disappeared from the country) but by the political revolution, war and imperial dominion of the 1799–1815 period and by the division of the commons. The former Bourbon forest law of 1755 (*De incisione arborum*) was praised as a good provision and the real culprit for the devastation of woods in the last half century was reputed to be the partitioning of the *demani* among the poor[253]. De Rivera thus reversed the narrative of disaster created by the previous generation: the 'disorder of water' was being caused not by a lack of private property but by its very introduction in a country where cultivable land lay mostly on hilly and mountain terrain, the plains being swampy and malarial. Priority lay with the reclamation of such swamps, so that agriculture could return to its 'natural' place – the plains. And the agency for re-establishing this natural order disturbed by Man was the State.

De Rivera's view of environmental policy was firmly Hegelian: only the State enjoys a comprehensive gaze over the country's physical nature and political economy, such that a general plan for the wise economy of water can be established. It is not up to private landowners to envision the effect of their actions over distant places, or to care about them. And even if they did, how could single disparate actions, he asked, be coordinated towards the common good? It behoves 'the beneficent hand of the King' to lead the citizens' actions towards both private and public utility by means of good laws and good administration[254].

The relevance of De Rivera's vision within the context of Neapolitan political economy of the time is the idea that public and private interest do not necessarily coincide and that private property should be regulated according to the nation's interest. His vision is not, however, of a return to common property. Rather, he distinguishes between false and true private property, the first being that created by the revolutionary laws for the partitioning of the commons, the second being that created through market transactions[255]. In this way, he maintains an ideal distinction between good property – that held in bourgeois hands; and bad property – that held by the poor, considering the latter a major cause of devastation through deforestation for subsistence purposes. Much less investigated in De Rivera is the nexus between bourgeois property and deforestation for market purposes[256].

The major contribution that De Rivera made to political economy ideas in the Two Sicilies was with a book published in 1833 under the flamboyant title *Considerazioni sui mezzi da restituire il valore proprio a' doni che ha la natura largamente conceduto al regno delle Due Sicilie* ['Considerations on How to Return Proper Value to the Gifts that Nature has Largely Granted the Kingdom of the Two Sicilies'][257]. The book was an investigation into the natural resources of southern Italy, conducted in the form of a travel report. While simply restating the accredited narrative of the decline of Magna Graecia, the original contribution of Engineer De Rivera consisted in articulating the political economy discourse along the lines of the country's watersheds. De Rivera's plan for restoring the true value of the gifts of nature was a highly complex scheme including many aspects of the nation's economy. Throughout his book, he promoted an idea of the civil engineer as the one who better than others could envision wise political economy, fit for the country's needs. After describing the country 'as it really was' (i.e. in its physicality) while also prospecting the possibilities for improvement, the second part of the book developed engineering schemes for redesigning it, through reforestation, scientific silviculture, reclamation, river channelling and a series of new harbours in order to improve navigation. To these De Rivera added plans for the regulation of internal and external commerce and road construction schemes. The author concluded with the proposal of creating a corporation of scientists and artists entrusted with designing and directing the nation's 'great restoration works'. A one-man work, the *Considerazioni* appears like a gigantic, all-comprehending and utopian plan of public works and economic policy at once. Unlike in Galanti's *Descrizione*, however, people very rarely figure in De Rivera's book and when they *do* it is as disturbing agents of the good natural order and transgressors of good laws.

The *Considerazioni* filled a lacuna in the Neapolitan school of political economy of the time, which paid very little attention to the huge question of the *bonifica*[258]. Improvement, for those authors, was ideally any agrarian investment following enclosure; in the physical reality of southern Italy, however, such investments had to tackle a whole situation of interconnected hydrological imbalance between mountain and plain. As such, improvement required what De Rivera called the 'benevolent hand of the government' to be planned first and then pursued.

Whether benevolent or not, the Bourbon government's hand lagged behind in both timing and largesse. Despite the attention that the 'disorder of water' had gained as an undisputed cause of economic backwardness and public unhappiness, discussions over the funding of reclamation schemes lasted some 25 years: the debate concerned the introduction of the institution of concession as a way to make land reclamation an entrepreneurial business. Both the scarcity of interested entrepreneurs, though, and the opposition of landowners led to the failure of the project. A decree in 1839 generically restated the current practice of *ratizzo* – that is, the apportionment of land reclamation expenses between the State and local landowners.

Despite the weakness and marginality of the country's entrepreneurial groups, Neapolitan economists of the Restoration intended private (and especially land) property as an undisputed good and a cause of public happiness. Economic progress was identified with the abolition of all customary practices of resource management, also considered responsible for environmental degradation[259]. It was precisely against this ideology that the notion of *bonifica* as 'general interest' and as a State prerogative clashed, leaving in its wake a weak and un-improving environmental policy. It should be noted, however, that by the late 1850s the public–private balance had shifted towards 'general interest' in both the Neapolitan political economy school and in public expenditure policies, such that a centralised general Bureau of Reclamation was finally created in 1855; unfortunately, the Bureau could only work for a few years before the Bourbon State collapsed[260].

***

The Economy of Water

In 1861, when a new State – the kingdom of Italy, governed by the Savoy King Victor Emanuel – took over from the former Bourbon State, the south was fully integrated into the European political economy as an agrarian periphery. Far from being an immobile social scene, the south had seen an agricultural and commercial revolution during the nineteenth century. Commons (be they woods, pastures, lakes or swamps) were being parcelled out and sold or given for rent to entrepreneurs; forests were being entrusted to private owners for the sake of their preservation; guilds had been abolished and manufacturing labour liberalised – and in some places, like the Liri Valley, also mechanised and disciplined in the factory; and not only land, but water too, was being appropriated, in order to fuel the new economy of vine and citrus plantations and power the hydraulic machinery of textile manufacturers.

Not even the rivers had escaped enclosure. A land-related resource, whose market value had increased tremendously through the expansion of a commercial economy, water was subjected to new, aggressive forms of appropriation during the nineteenth century. As a consequence, a new economy of water took shape in southern Italy's river valleys; it became embodied in the land and people of different places, producing new socio-ecological relationships and social costs. The Liri Valley was one of those places.

Chapter Four

Water Property and the State

It is the taking any part of what is common, and removing it out of the state nature leaves it in, which begins the property; without which the common is of no use. And the taking of this or that part, does not depend on the express consent of all the commoners.

J. Locke, *Second Treatise of Civil Government*

In the matter of property, use and abuse are necessarily indistinguishable.

P.J. Proudhon, *What is Property?*

The appropriation of water is probably the most relevant place to look at the birth of European industrial capitalism. In considering the history of industrialisation, we might say that industrial capitalism was born and raised in river valleys, where early entrepreneurs experimented with the appropriation of water and its transformation into waterpower, the primary engine of the factory system. The nineteenth century Liri Valley is one context where the transformation of water into property can be closely observed, along with its social and ecological outcomes. A situation of open access was created soon after the abolition of feudalism and especially with the incorporation of the Kingdom of Naples within the Napoleonic empire. As a consequence of liberalisation, the enclosure of water followed: it was a material process, carried out by fencing the river with stones and wood fascines, diverting water through millraces and into hydraulic engines which physically occupied the streambed, eventually producing a new, industrial riverscape. But water property – that is, legal entitlement to the possession of such enclosed parcels of watercourse – was also a discursive process. In other words, to be considered a legitimate practice, appropriating the river required the elaboration of a specific rhetoric of water property. The possibility for this discursive appropriation came from a particularly fluid definition of the relationship between public and private, such as that enshrined in the Two Sicilies' body of law.

This chapter will look at the process of water appropriation in the Liri Valley, trying to make sense of what it meant in terms of society–nature relationships. The historical evidence from this little valley at the periphery of the Industrial Revolution makes up a different picture from what is expected by 'tragedy of the commons' theories. As biologist Garrett Hardin has noted, once 'open to all', the river became an object of overexploitation and environmental degradation. Contrary to what is assumed in Hardin's theory, however, neither the enclosure of the river in individual properties nor State control over water were efficient responses to the Liri Valley's problem. Coming after decades of denigration and actual eradication of common property by the Kingdom of Naples, the enclosure of the river resulted in a true social and environmental disaster. Though apparently so favourable to industrialists, the 'liberation' of water in fact increased uncertainty and raised transaction costs – namely the cost of asserting and enforcing property rights[261].

Before being an environmental tragedy, the appropriation of the Liri River was a social one; and not only for the common people in the Valley, who were both forced to undergo factory discipline and periodically flooded. The tragedy, to begin with, concerned the appropriators themselves – the entrepreneurs. What follows is their part of the story.

Picture a River Open to All...

Issued in March 1817, the kingdom's Code of Public Law reproduced the Napoleonic Code's definition of State water as that which was navigable or apt for flotation, all other waters being 'public' – that is, *open to all* who could claim legal title to them. Such titles could be acquired through the purchase of properties (land and/or buildings) to which water rights were annexed, while permission for building new mills or enlarging the existing ones had to be requested from the Minister of the Interior (art. 21). The most common interpretation of the law became that which held that the property of water was linked to that of parcels of real estate – either land or edifice. This one article of law, however, was too little to be of much help in the massive upsurge of water litigation throughout the country. In fact, virtually no watercourse in the entire kingdom, except the Volturno, was navigable; also, most of them featured torrential regimes, being subject to exsiccation during the summer. This feature of the nation's rivers brought many plaintiffs to contest the very nature of watercourses, on which they claimed property, by defining them as mere torrents, not even perennial and thus not subject to State property. As a consequence, water disputes prompted a noteworthy portion of the juridical debate in the kingdom, some becoming famous for their political implications.

Probably the most important dispute, for contemporary debates on water property, was that between the Marquis of Sortino and the State over the property of the ancient aqueducts of Siracusa and the river Ánapo in Sicily. The Marquis, to whom the State had granted exclusive property of the old mills and waterworks that he had formerly enjoyed as a feudal possession, also claimed ownership of the ancient aqueducts of Siracusa, whose water moved his mills. He attempted to do so by stating that the river Ánapo was nothing other than a bundle of individual, non-perennial streams, each pertaining to lands of his own property. Two Neapolitan lawyers, Giuseppe Laghezza and Antonio Ranieri, were inspired by the case to write their *Breve ricapitolazione di una teorica legale intorno alle acque perenni* ['Brief Summary of a Legal Theory of Perennial Waters'], published in 1845. The lawsuit did not concern the right of use, the lawyers noted, but that of property, that is, the incontestable public nature of the Siracusa waters. If the Marquis's argument were to be accepted, the authors concluded, 'then the Danube, the Volga, the Ganges and all the greatest rivers on earth, each being no more than the sum of many, small and often imperceptible parts, would quickly disappear, or would become a number of private streams'; and with them would disappear the great nations to which they gave life and power. The book's very topic was thus the nature of property. The authors defined it as 'any private utility which is subject to a greater public utility'. In consequence of such a definition, the authors considered the appropriation of non-perennial watercourses possible and indeed desirable, because 'the occupation of their bed can be easily achieved and it would bring a public utility of immensely greater value than the private'[262]. Private property, in sum, was based on a dual terrain: nature (the non-perennial character of watercourses) and the public good. The comparable example was that of a man who occupies a wasteland to cultivate it: he would be more useful to the community than to himself said the authors.

This typically Lockean tale shows the extent to which the Neapolitan middle class shared classical political economy ideas. One of the two authors, Antonio Ranieri, was a well known lawyer in the capital city and a liberal *patriota* of the Italian Risorgimento – censored and even imprisoned for his anti-Catholicism. After the fall of the Bourbon State, he obtained the chair of Philosophy of History at the University of Naples and became a deputy in the Italian Parliament. Ranieri's extensive clientele and his network of political alliances was widespread throughout the kingdom; he practised both as a lawyer for the public estate (the *Demanio*) and privately, mainly for the provincial bourgeoisie. Especially versed in literature and history, he was well travelled throughout Italy and Europe and corresponded voluminously with liberals and literary figures in France, Tuscany and the Pontifical State. His views were representative of what may be termed the liberal public opinion of the time[263].

In Laghezza and Ranieri's theory of water property, public waters were not those allowing for transportation, as in the Napoleonic Code, but the perennial. This shift testifies to the differences between French hydrology (even that of Mediterranean France) and southern Italian. In the Two Sicilies, to claim public control over navigable rivers equalled claiming no public control at all over the country's waters. In fact, the redefinition of public rivers as being all perennial streams, pursued by the Crown between 1835 and 1859, implied the formal extension of State control over the nation's water. By that time, however, southern Italian rivers had already been subjected to primitive accumulation under the open access regime of previous decades. Rivers were not the same as before the Revolution, just as neither mountains nor plains were – nor people. This was even more evident in places such as the Liri Valley, where powerful economic change was occurring.

Coming after roughly thirty years of a regime of open access to rivers, that had caused both hydrological disorder and never-ending litigation throughout the kingdom, the 'Brief Summary' certainly did not contribute to the solution of such problems. Its main historical significance is that it represents a political discourse on water, expressing the interests, opinions and aspirations of the Neapolitan bourgeoisie, especially its provincial arm. In fact, members of that bourgeoisie largely resorted to lawyers like the authors to defend water rights against each other's claims, by invoking the 'public' nature of rivers. What this provincial middle class could more easily get, from Laghezza and Ranieri's treatise, was that: 1) water was free from either feudal or communal control; 2) the main targets of water law were former feudatories in their attempt at regaining old privileges; and 3) as 'public' property, non-perennial rivers (a great proportion of the country's water) were *res nullius*, i.e. open to all. Their best possible use, however, was to be appropriated by improvers and developed for the public benefit.

To decide what 'public benefit' really was, however, was no easy task. A decade later, summarising the results that the 1806 law had produced in the matter of water management, Ludovico Bianchini, a former Minister of Finance in the Bourbon State, would write that:

> Since that law had not made any rule [...], a great confusion arose and was further nourished by the various sentences of the Feudal Commission concerning water rights on various rivers. In several cases the Crown, on demands from some individuals, had given entitlements to make particular works on some river [...], but the efficacy of these directions had always been impeded by a great difficulty in defining if the work brought a public benefit, or if it was damaging other people's rights. Hence followed those stubborn and expensive suits before our courts, often decided by erroneous technical advice[264].

Despite the 'great confusion' reigning over the water of the kingdom, the culture of appropriation was justified by the commonly shared belief in the coincidence between private property and the public good. As one of the very few areas of the country where industrial transformation was giving water an unexpected market value, the Liri Valley played a crucial role in the development of the new water politics of the age of the Restoration. When, for example, an engineer from the Corps of Bridges and Roads was sent to report on the (assumed) threat to public safety posed by the De Ciantis Bros. wool-mill in Isola Liri in 1831, he expressed a very favourable opinion: the De Ciantis waterworks, the engineer claimed, 'being directed at improving their factory, benefit the public in every respect'.

A previous Town Council decision, however, had defined the De Ciantis mill as 'illegal and damaging to public health' – for it caused the flooding of a public road and the stagnation of water – and had ordered the destruction of the waterworks. But the decision had never been enforced. The reason lay in the opinion that the De Ciantis, owners of one of the five most important wool factories in the valley[265], also 'deserve[d] to be encouraged in the enterprise of their factory, which provide[d] a living to many workers indeed'. When the De Ciantis brothers, actually encouraged by such a statement, started to enlarge their waterworks, it was the owner of a downstream mill who opposed their initiative, requesting the *Intendente* to judge the environmental impact of the new works. The plaintiff based his accusations on both the arguments usually employed in legal disputes of the sort: 1) the mill threatened public health; 2) the De Ciantis had no right to the water they had appropriated. These claims, however, were promptly rejected by the State engineer, who declared the new De Ciantis' fence to be as 'necessary to the movement of their machines' as '*those machines [were] necessary to the Nation's culture*' (my emphasis)[266]. The engineer also claimed that it did not really matter if the De Ciantis had legal title to the water or not, since they *deserved* to have it. And they probably did get the permit to enlarge the factory, for their workforce doubled from the 150 of 1831 to the 300 of 1842[267].

The Appropriators

Until the end of the Kingdom of Naples, lawsuits reiterated the celebration of individual property rights over water. The Liri Valley was one of the areas where the language of appropriation was more openly played out, in the courts as well as in the field. As the *Intendente* of *Terra di Lavoro*, referring to Isola Liri, wrote to the Minister of the Interior in 1835:

> In that town, litigation over the use of water is frequent, since water moves so many machines, and it fuels the envy of many manufacturers, for the advantage and opportunity they have thus established[268].

The wool industrialist Gioacchino Manna, one of the first to move his workshop from Arpino to Isola in the 1810s, soon became the fulcrum of many, longstanding litigations involving the Town of Isola Liri and several other industrialists. Inherited by Gioacchino's successors, the Manna family's court cases – v. Carlo Marsella (from 1819 to 1823), v. Clemente Simoncelli, Carlo Lambert and Giuseppe Polsinelli (from 1834 to 1836), v. Giuseppe Courrier (1838), v. the Town of Isola Liri (in 1823 and again in 1866), v. the Coccoli family (from 1872 to 1883) – were still being brought to court almost a century later. They can thus be considered a good vantage point for what happened in the Liri Valley at the time of the water enclosures.

The first recorded case involved Carlo Marsella, also a wool industrialist, whose mill was located downstream from Manna's property. Starting in 1819, the latter wrote to the *Intendente* to contest the fence that his neighbour had built in the river-bed, also denouncing the Town Council for not having proceeded against

Manna. Since the fence had been built with no request for a permit, Marsella easily obtained a decision directing Manna to destroy his illegal constructions. The Minister of the Interior, for his part, promptly ordered the Town to issue a regulation for the partitioning of water among users, so as to have an instrument for the resolution of all future litigation. This recommendation, however, remained completely ignored. Meanwhile, Manna had turned to the *Corte dei Conti*, the highest court in the country, which annulled the *Intendente*'s decision: expressing a very liberal point of view, the Court argued in favour of Manna's free enterprise, on the basis of the fact that the Town had not contested the waterworks. Having failed to convince the Court that Manna's fence was damaging not only his downstream waterworks, but also public health, Marsella finally signed a private agreement with his rival, by which he obtained monetary compensation for the damage. In exchange, he did not pursue further action before the Council of State[269].

Soon after, however, a major lawsuit started between Manna himself and the Town, concerning a ditch that the industrialist had dug on one of his estates named San Sebastiano, immediately adjacent to the former Ducal Palace[270]. The property included a number of buildings endowed with water-plug and hydraulic machines, formerly used by the Duke as copper and silk mills. The new ditch was intended to increase the waterpower available to the mills, which Manna had incorporated into his wool factory. The San Sebastiano property was cut across by a country road, called Valcatojo. On its way down from the river to the factory, the new ditch had to cross this road. Trouble began when a group of neighbouring landowners claimed that Manna's ditch had the effect of 'drowning' the Valcatojo road; the Town Council – whose Mayor, Luigi Spagnoli, was also one of the plaintiffs – ordered a 'restoration to pristine'. Manna responded through recourse to the *Intendente*, claiming his right to do whatever he wished on his own property; in turn, he accused the plaintiffs of having caused major flooding of a public road, called La Selva, by their own irrigation ditches. The La Selva road being now impassable, Manna argued, people needed to use the Valcatojo, which followed roughly the same direction. In the subsequent suit, the Town called local people to testify to the public character of the Valcatojo, used by locals 'from time immemorial'; meanwhile, civil officials and private experts searched the site looking for material evidence of the public or private character of the road. Both witnesses and experts were called to certify a right of property which, evidently, had never been endorsed. The one certain fact to emerge from witnesses' statements, however, was that the La Selva road was really flooded and had become impassable.

To complicate things, the Valcatojo road was apparently the only way to get to a small chapel located at the top of the San Sebastiano hill. The chapel, formerly pertaining to the Duke's palace, was now Manna's property. Although Manna had never obstructed passage to the chapel, he threatened to do so when he felt his rights under attack. Neither the property of the road nor that of the chapel, moreover, were of any major concern for the local people, who apparently climbed up to the place only once a year, in procession, for the festival of the Assumption. In August 1835, outraged by the opposition that the Town Council still maintained against his waterworks, Manna tried to prevent the procession to the San Sebastiano chapel from trespassing on his property. Alerted by the local police, the Intendente authorised the dispatch of gendarmerie in order to 'prevent riots' and 'save order and concord'[271].

The tradition of the procession, however, had only recently been invented: not more than fifteen years ago, according to the 'ancients' of the Town. The coincidence is striking between this date and that on which Manna had come down from Arpino with the purpose of mechanising his workshop. One might suspect that some envious individuals or rivals had plotted to originate the procession, in order to claim public control over Manna's property. Even without resorting to such conspiracy theories, the core issue was that, although Manna enjoyed full ownership of his property, he also had to allow public passage on the Valcatojo road. Documents attesting to the existence of this obligation, however, were nowhere to be found. The reason is that, before the San Sebastiano hill became private property, it was an un-enclosed part of the feudal domain of the Duchy of Sora. As the older inhabitants of the Town reiterated, the Duke had sold all of his domain to the King and the King had sold the San Sebastiano property to Manna: thus, a certain councillor Villa observed, the Town had better not sue. In the process of emerging from feudalism, the hill had been enclosed and had become property.

The kingdom's Civil Law limited the free enjoyment of private property with the prohibition of damage to others, or to the public. Since Manna's ditch only ran through his own property, the plaintiffs had to resort to the discourse of the public and to Public Law. This, as we have seen, was particularly unhelpful when it came to water suits. In fact, although the *Intendente* directed all landowners to repair flooded roads, whatever their ownership, and prescribed that a permanent irrigation canal be built at the expense of the users, the plaintiffs could take their case to the *Corte dei Conti*. As in the previous example, the Court expressed a different opinion from that of the *Intendente*, declaring him incompetent on the issue, which was reputed to be a matter of private rights. What was at stake, in reality, was the property of water. If the *Intendente*'s decision was to be enforced, this would transform local landowners into users of a common infrastructure, the irrigation canal. But the *Corte dei Conti* came to the rescue of the plaintiffs, who were granted the unquestionable right of freely enjoying the water, while retaining the right to enter into private disputes. This ruling clearly did not resolve the issue: it only prevented plaintiffs from calling on the State in their litigation about the Valcatojo road. In fact, water conflicts in the valley continued apace for more than a century afterwards.

Given the strongly localised character of industrialisation in the Liri Valley, personal rivalries evidently played an important role in disputes such as the above. Not all of them took place in Isola, however. One major lawsuit, for example, concerned the Picano brothers and the village of Sant'Elia on the Rapido River, a confluent of the Liri a few miles downstream from Sora. In April 1835, the Picani wrote to the Minister of the Interior to ask that their 'right to use water not be constrained by local interests'. Wanting to mechanise their wool factory, a few months before they had asked the Village Council for permission to build a number of mills on a property they owned adjacent to the Rapido, in order to move new machines they had imported from France; but the Mayor, Francesco Lanni – also the owner of a wool-mill in the same village – had prevented the Council from discussing their request[272]. The Minister soon instructed that the assembly be presided over by a district councillor. Although the latter only came from the nearby Sora[273], it took three more months for the Picano affair to be discussed by the village Council, which eventually found nothing to say against the request[274], but only directed that two bridges be built by the Picano in order to preserve two country roads nearby from possible water damage. Soon after, however, the Council sent a petition to the Minister, claiming it had 'been mistaken' in its previous decision and that the requested waterworks would bring a serious threat to public health and to 'the interests of those inhabitants'. Signed by a number of villagers, including three members of the Lanni family (Francesco, the Mayor; Giuseppe, the councillor; and Raffaele, the clergyman), the letter explained that the Picani were actually constructing works on a far larger scale than they had claimed, which would completely divert the river course, drying up the existing bed and thus depriving the local population of 'an infinity of advantages': trout fishing, cattle watering, waste disposal and riverside areas where grain and clothes could be laid out to be sun-dried. Besides, the Picani's new machines would also eliminate a number of jobs. Finally, public health was at stake: by narrowing the river-bed at certain points, the projected waterworks would increase the power of flowing water and so too the risk of flooding[275].

The Picani promptly responded that the petition had been prepared by 'one our particular enemy' – Lanni – and signed by his workers. The Mayor, in fact, was the owner of 'multiple waterworks' on the Rapido River: a paper-mill, a wool-mill, a grind-mill, an olive-mill and a textile mill, all located upstream from the Picano property. The latter thus asked for the issue to be examined by the parties' experts, who were later appointed by the *Intendente*: one for the Picano, one for the Village and a third to represent the *Intendenza* (Regional Government) – and thus the State[276]. By appointing a representative of the 'public' interest, the *Intendente* implicitly admitted that this differed from that of the village, acknowledging the *de facto* control that the Lanni family exercised over the local polity. Thus, despite this being clearly a dispute between two manufacturers, the *Intendente* could not but frame it as one between the public and the private: he thus asked the architects to indicate measures to 'reconcile the protection due to a Manufacturer and the interests of the inhabitants'[277].

The inspection produced a piece of truth that had not yet been mentioned in the documentation: it emerged that the Lanni family owned the only fulling machines in the area, so the Picani had to bring their wool to be fulled there, at huge annual expense (900 ducats)[278]. By building their own hydraulic machines, the Picani would not only have saved money; they would also have broken the *de facto* exclusive control over waterpower enjoyed by the Lanni family.

Fortunately for the Picani, the Council itself was only willing to support the Mayor up to a point and this point had a price. When the first bill of 24 ducats had to be paid by the village for legal expenses, councillors deliberated to settle. The Picani were accorded the permit to build their machines subject to their compliance with a number of technical conditions in order to prevent future damage to third parties and the village[279]. Conversely, the price in money of freedom for the Picani was much higher: besides committing themselves to pay for all the requested modifications to their project and to reimburse the Village by paying an annual fee for the use of water, they agreed to pay for all litigation and inspection expenses (over 700 ducats). They had finally won their 'freedom to use water', but at what a cost!

Like the Manna case mentioned above, this lawsuit also tells us something relevant beyond the micro level, something pertaining to the sphere of culture. The dispute can be read as a discourse (what else are lawsuits, after all?) on water property and rights. Each party's lawyers had hotly debated, one in favour, the other against, the existence of an exclusive right of property held by the Picani on the water flowing through their property. As in the Manna case, water property was linked to that of the land. As a consequence, both lawyers' arguments hinged on the ownership of a tiny plot lying on the other side of the river facing the Picano property. Did that plot belong to them, too? If yes, that stretch of river-bed would, in legal terms, be private property and so too the water flowing through it. While writing their treatises on this Byzantine issue, both lawyers neglected to mention that the Picani had never claimed property over that water: in fact, they had merely asked for permission to use it[280].

From the Picano story, one can also infer that litigants in the Liri Valley became further convinced of the importance of claiming exclusive property over water, by any means necessary. In fact, lawsuits often followed after violent action had been taken, such as dam-breaking or illegal construction of waterworks on third party properties. In 1834, for example, the mill-owner Clemente Simoncelli from Sora wrote to the Minister of the Interior to complain of 'assaults, acts of violence and abuses' that his rivals (Manna, Giuseppe Polsinelli and Giuseppe Courrier) had committed against the new hydraulic engines he had placed along the river.

His 'enemies' – so he complained – had also opened a plug and diverted waters upstream of his mill, so he had brought a suit against them at the court of primary jurisdiction. They had destroyed a fence on his property, stopped three corn-mills, two fulling-mills and the machines in his wool-mill, 'whose soul', Simoncelli wrote, 'is none other than the water'. Another suit had been brought, but again the following spring one of his rivals 'had armed his workers' and built a wall to keep Simoncelli from pumping water through some of his pipes on another tract of the river. A riot was quelled by the gendarmerie and another lawsuit followed. In the meantime, another of his rivals had called on the administrative powers to judge against a new mill that Simoncelli had built to continue the processing of wool. Simoncelli thought that the real aim of this move was to weaken his business by making him waste money in legal expenses. Nonetheless, he filed a suit against another mill-owner who had built waterworks on his property. In the aftermath, he wrote to the Minister of the Interior that a decisive intervention by the State was needed, since 'only the arm of the law can restrain such arrogance'[281].

A few years later, it was one of Simoncelli's rivals, Courrier, who reported to the *Intendente* how Manna had destroyed part of his waterworks, while the *Intendente* recommended that both litigants 'restrain their actions within the due limits'[282].

An emblematic case of violent action concerned the wool industrialist Giovan Battista De Ciantis and the monastery of Santa Restituta. On the night of 29 to 30 October 1839, De Ciantis demolished the temporary dam (a palisade made of fascines) belonging to the monastery on a tract of the Liri upstream from the dam driving his mill. By his own admission, De Ciantis did this because, some years before, the monks had put up the palisade and this prevented water flowing to his mill. In the ensuing suit, heard by the royal judge of Sora, the clergyman Lanna declared that the Monastery had acted within its rights, for 'as is well known, such fascine-dams have to be remade every year, to prevent them from being destroyed by winter floods'. In order to exercise this right, however, the clergyman had destroyed De Ciantis's water plug, which allowed the excess water from the monks' mill to reach his property. Needless to say, the judge ruled in favour of De Ciantis, since the latter had been 'disturbed in the possession of his watercourse'[283].

Despite its verbal reiteration, the right of water property in the kingdom was anything but certain: in fact, it was a highly contested practice. In the end, the private appropriation of water resulted from a situation of uncertain transition and experimentation, during which violence and illegality proliferated and were even tolerated, to some extent. Perhaps the best description of this situation is that given in 1857 by the wool industrialist Felice Viscogliosi from Isola, who happened to build his mill-dam after a Royal Decree of 1853 claimed that all running water in the country belonged to the State. Too bad, Viscogliosi lamented, that only his waterworks were now incriminated as abusive, while many others existed along the same stretch of the river[284]. His remarks clearly achieved their aim – that he could himself benefit from the same tolerance of private abuse in the name of the public interest – for the *Intendente* asked the Minister of the Interior to suspend the sentence against Viscogliosi and to issue a 'very exceptional economic measure' to save his mill. The reason was that Viscogliosi, after all, deserved the same regard that had been granted to other manufacturers, the 'tacit and long-lasting consent' accorded to their costly works and highly valuable factories[285].

The royal decree of 1853 was not the first that aimed at ending water disputes by defining water as State property: two other decrees, in 1839 and in 1850, had already been issued with the same intent, but the problem remained that no prescription for administering these 'public' waters was included within the Law of Public Administration, compiled in 1816. Actually, the law reflected Minister Zurlo's 1809 instruction that 'the partitioning and use of public water, including public dikes' was a matter for local government and was to be managed through what the law called the 'Regulations of the Rural Police'. These were aimed at 'the health, safety and preservation of the countryside, of its animals, tools and products'. Town Councils were to vote on these regulations and then submit them to the *Intendente* for his endorsement. The regulations of the rural police could impose pecuniary fines for damages or a custodial sentence of up to three days, while the law expressly forbade private citizens to maintain armed bands to defend their property.

The problem with such water politics was that neither town councils nor local landowners (mostly the same people) were interested in subjecting 'public' water to any regulation; in fact, they preferred a situation in which water property went along with land property and water-flows could be held in exclusive ownership like land. The liberation of water meant to them that rivers were now open to anyone's appropriation: this resembled a colonial annexation, where individuals were allowed to enclose land and natural resources that 'their' Crown had claimed to itself, after wresting it from previous indigenous–feudal–common property regimes. We might call this appropriation of water a 'primitive accumulation': not only were water rights exchanged on the market, bought and sold like any other asset – along with the old monasteries and feudal mills to which they belonged; but rivers were also materially appropriated, by fencing water with stones and fascines, or destroying someone else's fence. A generation of industrial entrepreneurs, for whom feudalism was a story with a happy ending told by their fathers, was now experimenting with the possibilities opened up in these 'liberated' places. If the world was no longer the same as before Napoleon (and Arkwright), still less was the natural world. Just as the Industrial Revolution had made water a key resource to progress, so the French Revolution had opened up rivers to the not-so-peaceful conquest of capitalists.

Water Property and the State

Water Wars, Water Discipline

If privatisation was not a good response to litigation costs arising from the open access regime, neither was State control. This can be argued from what happened after southern Italy was incorporated into the new Italian nation.

As regards rivers, there was an important difference between the new State and the old. Public waters were included within the Public Administration Law, issued in 1865, whereby the King of Italy, Victor Emanuel II of Savoy, claimed actual control over the nation's territory, population and natural resources. Although the Bourbon King had already declared all watercourses to be State property, this principle had never been enforced; the Savoy State, on the other hand, pursued its aim to control the nation's water. Italian rivers were not open to all: they were State property; their use was granted by the Government, regulated by Prefects and subject to fees; the income they generated was subject to taxation. Current 'owners' of public waters had to legitimise their position by asking for legal permission to use 'their' water. They had to comply with multiple *Regolamenti Disciplinari* [Disciplinary Rules] following legislative change over the years.

Nevertheless, disciplining water use ended up being a much more drawn-out and more complicated undertaking than expected. In the end, the property scenario remained unchanged. In fact, the 1865 law recognised water rights granted under the previous regime, whether they were based on legal titles acquired thirty years before the new law, or on customary uses practised 'from time immemorial'. As under the previous regime, the ideology of economic progress demanded that 'tacit consent' be granted to industrial capitalists, whose vision of water rights was only apparently threatened by the new State.

Not even this compromise, however, succeeded in ending the water wars. When, for example, the mill-owner Andrea Tuzj of Sora asked the Prefect for permission to build a canal to take water from the Liri, he had to face the opposition of the legal 'owner' of that same water, the wool industrialist Vincenzo De Ciantis. The latter invoked his 'peaceful possession from time immemorial and exclusive domain' over the same tract of the Liri, 'that the late King had granted him by royal decree on 5 August 1838'. A concession to Tuzj would be a real 'usurpation', the opponent claimed, not only damaging himself, but also the people of four villages who used the same water for irrigation, as well as the public lavatory of the Town of Sora. Although the water flowing through that stretch of the river was his property, De Ciantis said, he in fact had consented to share it with the population and with irrigators, 'in order to make agriculture prosper'. Exactly for this reason, he could not countenance 'that Mr. Tuzj should usurp a water plug without any right or title'[286].

In the water wars of the post-unification era, litigants mainly struggled for legal recognition of acquired water rights; but they also profited from a situation of institutional shift and uncertainty, trying to increase the permitted extension of their 'properties'. A major case was that between the industrialists Felice Viscogliosi and Francesco Roessinger between 1869 and 1876. What the more than seven volumes of judiciary documents concerning this case tell us is the epitome of stories of water litigation in the Liri Valley, one that contains the rhetorical instruments commonly employed in a variety of other cases over many years. It all began when Roessinger asked the Prefect's permission to enlarge his water works on the Liri River and Viscogliosi, the owner of an upstream property, opposed the request for the projected works, arguing that they would threaten to damage his mill. The Prefect, however, did not consider Viscogliosi legally able to mount a challenge, since he had not yet legitimised his own water plug. While the latter responded that his waterworks preceded the royal decree of 1853 and thus had to be considered private property, the Prefect claimed that the Liri, as a 'navigable' watercourse, was a public river even before that decree and that Viscogliosi's water plugs – lacking any formal permission – were illegal. From a dispute between individuals over water use, the issue then became one of defining the 'public' or 'private' character of the river. The person of the Prefect himself was drawn into the dispute. Viscogliosi, in fact, sued the latter for being incompetent in the 'private' litigation between Roessinger and himself. In a surprising ruling, contradicting the legislative shift of the new regime towards the public nature of all Italian watercourses, the *Corte d'Appello* in Naples upheld Viscogliosi's right.

As in the G.B. De Ciantis case above, a key role was played in the dispute by 'the ancient Roman principle that the error, when general, gives birth to the rule' – a principle invoked by Viscogliosi in his defence. The banks of the Liri and Fibreno, the industrialist remarked, teemed with mills lacking any permit whatsoever: thus, 'either past laws did not prescribe such permits, or they are all illegal'[287]. At stake was no less than the entire factory system of the Liri Valley. When the case eventually arrived before the Council of State, this body insisted that there was no difference between the Bourbon and the Italian State's water law. Not only had the royal decree of 1853 – the Court noted – claimed all watercourses as State property, but the Zurlo Memo of 1809, by abolishing any privilege over rivers, had already made all water disputes a 'public' affair.

The latter Court decision shows how, despite the politico-institutional shift, great confusion still reigned over the country's water. But, as under the previous regime, the legal definition of rivers had very significant consequences for industrialists: Viscogliosi, for example, was firmly intent on winning the case because – as his lawyer declared in 1872 – 'for three years his factory has worked with half the waterpower, incurring very serious losses'. The industrialist thus claimed he would not 'stop fighting for justice' and threatened to go to the press, ask for a parliamentary enquiry, or pursue other lawsuits[288].

When, one year later, Viscogliosi agreed to ask the Prefect to legitimise his waterworks, his request was opposed not only by Roessinger, but also by Dionisio

Courrier, the owner of a paper-mill located upstream from Viscogliosi's property. In December 1873, an expert from the Office of Civil Engineering was sent to draw up the *Disciplinare* for the legitimisation of both permits: after two days of discussion 'in contradictory', during which the attempt to find 'any mediation' failed, the two industrialists asked for more time 'to understand each other and find an agreement'. One and a half years later, while this agreement was still being sought, Viscogliosi asked the Prefect for a list of the documents he himself had filed the previous year, 'in order to add a few, sober remarks' in response to those already added by his rival. When, finally, the case reached the Minister for Public Works (the competent authority for water permits) and the latter demanded that both industrialists conduct 'various and very delicate experiments' concerning the safety of their waterworks and an estimate of the amount of water to be paid for, before obtaining definitive legitimisation, both Viscogliosi and Roessinger once again sued the Prefect, contesting his competence on the entire issue and thus claiming, again, the private character of their water use.

This claim notwithstanding, Roessinger also sent the Minister a printed report concerning the 'present and previous state of the Liri water in the area concerned', which the Minister forwarded to the local Office of Civil Engineering – already overwhelmed by the documentation pertaining to this case. Clearly enough, both litigants pursued the strategy of protracting the legitimisation process for as long as they possibly could, while also periodically trying to reassert their property rights over the water, so as to postpone their final submission to the *Disciplinare*. What was at stake, again, was the ownership of water: if it belonged to the State, they not only had to pay for the use of the previously 'free' waterpower, but they also had to face costly modifications to their waterworks in order to make them acceptable to the State's technical office. Moreover, the concession would only last thirty years. To people such as the Liri Valley industrialists – used to considering the river as their own, enclosed property – the new, State property regime was particularly hard to accept, so they fought by every means in their power to evade it.

As under the previous regime, personal rivalries and entrepreneurial competition were also fought over the public–private nature of rivers. As the Civil Engineer remarked, this case was also imbued with 'the antagonism and long-standing acrimony between two entrepreneurs'. This had become strikingly evident when, in 1870, Viscogliosi had regularly requested a water permit for the Fibreno River, in order to compensate for the losses his mill was incurring as a result of the litigation over the Liri waterworks. His claim was soon opposed by Roessinger, who did not have any waterworks on the Fibreno River, but nevertheless tried to confuse 'rights and facts concerning the one case with those concerning the other'[289]. Two years later, the Chief Civil Engineer lamented the difficulty of deciding the Fibreno concession due to the 'excessive ease with which this Office accepted the oppositions, this being a case concerning two powerful individuals'.[290]

Water Wars, Water Discipline

The industrialists' resistance to the 'disciplining' of waterworks and their discursive attempt at reiterating the private nature of rivers are the main characteristics of the Liri Valley water wars in the second half of the century. In 1872, for example, Francesco and Vincenzo Manna (Gioacchino's sons) eventually sold part of the water flowing out of their canal to the Coccoli Bros, thus allowing them to open a new wool mill downstream from their own. By means of this unusual transaction, they aimed at demonstrating their exclusive and full ownership of the water they used. In fact, by evoking the possession of property titles 'from time immemorial', they repeatedly refused to undertake the legitimisation of their waterworks, requested by the Minister of Finance[291].

While neither private property nor the State could discipline water use in the Liri Valley, another agency – corporate capitalism – was effectively disciplining water, by accomplishing the total drainage of the great lake of Fucino, located roughly forty kilometres uphill, on the mountains of Abruzzo. Starting in 1861, the reclamation works were driving vast quantities of water into the Liri, through the Torlonia canal[292]; on the completion of works in 1874, the Liri River had gained an additional flow of 2.339 cubic metres, a substantial addition to the 6.7 cubic metres that the river measured at Isola Liri[293].

These changes in the river flow must have allowed for an increase in waterpower. Nonetheless, they also increased the risk of overflow: in 1872, the paper industrialist Dionisio Courrier asked permission to build a new concrete dam, claiming it would merely replace the two temporary ones destroyed by a flood in 1864. Called upon to advise on the request, the Civil Engineer responded that, due to the Fucino reclamation works, the 'state of the place' could not be compared to the previous situation of the river. Draining the Fucino had particularly influenced the height and power of floods and this fact, the engineer continued, 'deserves the greatest attention, considering the extraordinary damage which has since occurred'[294]. The Fucino project, therefore, prejudiced the definition of Courrier's water rights, acquired by his family in 1824, for it was 'impossible to ascertain the state of the river-bed and the height of the destroyed dams before the flood of 1864'. Not only the 'state of the place', but also the precarious equilibrium of relationships between riparian owners had been compromised by the Fucino project. These relationships were based on a private agreement that, thirty years before, Dionisio's father Giuseppe had signed with the neighbouring families: Silvestri, Belmonte, De Vito Piscicelli and Mazzetti Marsella. Dionisio's request to build the new dam was considered by his neighbours as breaking the old agreement. The industrialist defended himself by remarking that the Fucino works had caused not a momentary flood, but permanent changes in the structure of the place, 'a constant and daily fact which has transformed the situation of the river entirely'[295].

When, in 1898, Count Luigi Gaetani of Laurenzana purchased the former *Ducale* Palace, along with all the pertaining water rights and hydraulic works, and presented his request for legitimisation of water use, he found a very complex situation: the draining of Fucino lake had been completed and this had substantially altered the pre-existing adaptation equilibriums between water users roughly forty kilometres downstream. As a result, the Count's request had to face a five-year-long bureaucratic process, during which the Civil Engineer legitimised five out of the existing seven water-plugs, thereby lowering the amount of water-flow pertaining to the Laurenzana property. None of the increase in waterpower enjoyed by the Palace mills, due to the increase in the Liri water-flow after the Fucino draining, was 'legitimised' by the Civil Engineer.

Obviously, the Count did not easily give up on this matter of the loss of such a substantial amount of waterpower flowing through the two non-legitimated plugs. A protracted lawsuit started between Laurenzana and the Italian State, during which the Count contested the very idea of measuring the amount of water pertaining to the property, for such measurement was arbitrary in respect to the date when the property had been acquired. At the time when the palace's waterworks were created, he claimed, it was impossible to calculate the waterpower flowing through them; in any case, the law then did not demand such calculation. So things had to remain as they were, he concluded, if the current Italian laws acknowledged the water rights acquired by possession for over thirty years.

Measuring the waterpower of the Ducal palace, however, was also a problem of a hydrological nature. The Liri river flow, the civil engineers wrote, varied between 1.7 and four cubic metres, with an average of two, and even the flow from the Fucino lake varied between two and five cubic metres. On such variable flows of waterpower depended a number of downstream users, whose machines were moved by the water flowing out of the Palace's canals. As a matter of fact, the Gaetani di Laurenzana v the Italian State case involved some of the major local industrialists of the time, plus the Town of Isola and the neighbouring mill-owners[296]. All of them asked to participate, with their own experts, in the Civil Engineer's inspections. In addition, the industrialists contested the estimate that the Count had given of the water-flow resulting from the Palace's canal (seven cubic metres), claiming it was actually much smaller. The measurement of Gaetani's water-flow was thus a crucial issue for them: a wrong calculation would mean financial ruin[297].

We should not imagine, however, that local industrialists and landowners banded together against the State in order to save their cumulative access rights to the maximum amount of water-flow; as usual, in fact, the litigation also pitted individuals against each other and was fought by means of both public *and* private law. And the same happened, in those very years, along the riverbank of the Fibreno River. Another lawsuit involving a number of industrialists, and also hinging on the uncertainty of both water rights and water measurement, concerned the apportionment of the river-flow among users in Carnello. On the morning of May 18, 1896 – as all the industrialists were summoned by the Prefect to assist two government engineers in filing a report – the main act of the tragedy of water enclosures was performed on the open stage of the Fibreno riverbanks.

The Tragedy of Enclosure

The 'prologue' to this performance was a Government Decree, issued in 1884, ordering the compilation of a national list of public waters and their users. Only ten years later, however, did the Government issue a general Disciplinary Rule to make the law effective and help Prefects to complete the task. But the *Disciplinare* of public water simply asked all users to register their water plugs at the local office of Civil Engineering. Industrialists reacted to the new law in the usual way: they did not feel compelled to register their water rights, since these had been acquired more than thirty years prior to the new law. Without individual users' willingness to cooperate, the State had no real control over the situation of national waters. Nor were the Prefects particularly zealous in enforcing the new law, either because they knew all too well how users would react, or because they actually agreed that the use of water should remain free for industrialists.

This apparent equilibrium of institutional non-compliance was nevertheless periodically disturbed by a combination of human and natural agencies. This is what happened in Carnello in 1896.

The industrialists Courrier, Ciccodicola, Roessinger, Viscogliosi, Gemmiti and Società delle Cartiere Meridionali sued their rivals Lefebvre and Zino for damages and asked the Prefect to verify the legal status of their water plugs. It was thus by private initiative, not by a State attempt to gain control over the Fibreno River, that the Carnello litigation was started. Responding to the industrialists' request, the Prefect sent two civil engineers to ascertain 'what the legitimate uses, what the quantity of water used, whether the existing water works match the legally required criteria and whether they are proportional to the legitimate uses to which each is entitled, without damaging other users'. Clearly, such a task could not but involve all water users in Carnello. The Prefect, in fact, also ordered that the plaintiffs' waterworks be verified, to make sure they had legal title to stand in the controversy. A long and complex affair, involving technically difficult measurements and an impressive amount of legal documentation, the Carnello case gave the State a precious chance eventually to discipline all water uses in the area[298].

Thus, on the morning of May 18, a crowd of water users, each accompanied by an expert and a lawyer, gathered on the Fibreno riverbank near the Zino factory and the site visit of the engineers Mezzacapo and Silvestrini began. Soon after the first formalities, the Zino family lawyer stepped up to claim that his clients were 'not the owners of temporary concessions, but of a right of domain over the Fibreno, descending from purchased titles as well as from ancient possession and use'. As a result, they invoked the competence of private v. public law. Lefebvre's lawyers, for their part, followed a similar line of defence and from then on the affair was played out in the language of private right.

Despite opening in this totally unsurprising way, the meeting revealed how entrepreneurs also aspired to, and to some extent depended upon, the existence of some higher power, capable of settling their water disputes according to universal criteria. All the parties, in fact, asked for verification of the state of water use to be carried out, in order to ascertain not only the legality of that use, but also the quantity of water to which each user was entitled. Somehow, the Liri Valley industrialists conveyed the idea that the sum of private water amounted to a public matter and that the State was responsible for governing not water itself, but the relationships between individual owners.

At the same time, private owners aspired to some technical solution to their tragedy. They could see how the river was a limited physical entity, an open pasture – to use Hardin's term – where each user's action influenced the quantity and quality of water available to the others. After many years' experience of industrialisation, they were fully aware of physical interdependencies within the river-bed; in fact, their problem lay in finding a clear definition of each other's rights – in partitioning the pasture of the river. What they expected from the engineers was an exact measurement of the quantity of water to which each user was legally and materially entitled, considering the situation of the river and that of the existing waterworks. Lefebvre and Zino, for example, asked the engineers to give technical advice on a private agreement their predecessors had signed in 1848 for the management of a small dam located in Carnello. The agreement, half a century old, had been repeatedly contested by all the parties; nevertheless, they hoped that the engineers would be able to clarify which changes in the physical state of the river were to be ascribed to human and which to natural agency, or extra-local causes. Aware as they were that the river they faced was not the same as that of their forebears, they aimed to have the current situation recorded in official documents that might be referred to in the eventuality of future litigation.

For the engineers, in turn, the exact measurement of the flow was a means of overcoming the uncertainty produced by legal rights deriving from older times, in which waterpower was an undetermined entity and not the sum of individual hydraulic horses; overcoming uncertainty, in turn, was a means to end the very possibility of abuses. For the Government, the measurement of water was a precondition for establishing actual control over the country's rivers, without reversing old local equilibriums. The State's authority over water was to be based on the scientific method of hydraulic engineering, as opposed to the legal rhetoric mostly preferred by local users. In this way, old property titles were not formally contested: they were rather measured, limited and brought under State control.

Measuring rivers, however, was still a very complicated business for hydraulic engineering. A mathematically objective knowledge of rivers, expressed in the

The Tragedy of Enclosure

language of hydrodynamics, was at the time desperately wanted – and not only by the Liri Valley industrialists – as a prerequisite for a more efficient mechanisation of labour, as well as for disciplining both water and water users[299].

In addition, measuring the flow of a river so thoroughly used and transformed in the course of past decades was an almost impossible undertaking. After having determined, with great difficulty, the height of the plugs and that of the water surface at different points of the river-bed and in the dikes, and finally the flow of water in the respective tracts, the two engineers had to admit that the water surface on the third day read as being five centimetres higher than on the second and that variations were frequent within the same day, due to the intermittent use of the engines[300]. No technical solution was apparently available to waterpower users in Carnello. Probably for this reason, they had to periodically interrupt the engineers' work in order to corroborate measurements against information available from the respective title deeds and lawyers' reports. The appeal to written sources, in fact, was the main anthropological feature of water users' activities in the nineteenth century Liri Valley, whereby a legal, fictitious river was continuously superimposed on the real.

When the engineers paid their third visit to Carnello, the following June 18, things simply followed the expected script. Lefebvre's lawyers opposed Ciccodicola's request for legitimisation, for their client had been the first to present a similar request regarding the same water; Ciccodicola then said he would even be willing to pay the fee, 'if this [was] the reason why Lefebvre's demand should be preferred'[301]. The latter also contested the statement filed by Courrier's lawyers against the engineers' report and threatened to file new petitions and present new documents in order to demonstrate his rights. Ciccodicola then pre-announced his counter-statement to any documents his rival might eventually present. Zino's lawyer, for his part, insisted on the perfect legality of his client's property rights and full domain over the river. Gemmiti, the only woman actively participating in the controversy, presented her lawyers' statements, based on an agreement signed by her neighbours in 1876, according to which her mill was granted a fixed amount of waterpower, even during the dry season. She insisted on asking that the agreement be respected, even if it ended up being inadequate for the actually available water-flow and 'causing a shortfall in the waterflow she enjoyed at the time and a reduction in' the effective capacity of her own engines. As her lawyer stated, 'it could not be asked of Signora Gemmiti that she replace the engines with newer ones, both because she can use her water rights as she pleases and because such cost has not been imposed on her by the State's permit nor by the private agreement'[302]. It emerged that, after 1876, the Minister for Public Works had granted water permits for a total amount greater than that actually available in the river, thus leaving Gemmiti's mill 'completely dry'. While rejecting the available technical solution of setting up new, more efficient engines, the lawyers opted for a monetary one: they suggested billing the water users upstream for every un-worked day at the Gemmiti mill.

At the end of this third day, completely overwhelmed by the lawyers, the two engineers eventually surrendered. While mentioning that the rules would forbid them to do so, they agreed to allow petitioners to file further statements and/or documents by a certain date and any eventual counter-statements by a later date. They decided to do so in order to make sure that no more papers would be presented to them during their next inspection. On this announcement, the meeting ended in general agreement and the curtain can fall on our play. We can now see the industrialists in Carnello leaving the scene to go back to the family archives, or to compose other legal documents. The end of the story was still out of sight and similar acts of the same play would be performed for twenty years to come. The impossible task of enclosing watercourses on individual properties kept people engaged for a long time still, until hydro-electric plants came to replace the old machines in the river and water was enclosed in completely different 'networks of power'.

***

The story of water property in the Liri Valley challenges some of the basic assumptions in the account of the 'tragedy of the commons'. In the Liri, privatisation of water resources *did* occur as a reaction to a previous feudal–communal regime, but it *did not* produce efficient resource management in either ecological or economic terms. Exclusive property rights were claimed over water, but both the transaction costs and the environmental costs of industrialisation were persistently high throughout the nineteenth century. In the long run, as Garrett Hardin observed, free-riding brings ruin to all. While Hardin associated free-riding with the commons, however, it should more properly be associated with private property: in the Liri Valley case, for example, entrepreneurs acted as private owners of the water, imagining it as *natural capital* on which they could themselves free-ride. The 'tragedy of the commons', from this perspective, can be seen as a tragedy of water 'enclosure' and privatisation.

The Liri–Fibreno basin was not, in fact, an open access system – a 'pasture open to all' – but a common resource (the river) on which a basic form of exclusion had been established, without setting up any kind of common property regime. As a number of studies on the commons have shown, an important incentive for individuals to cooperate is the clear perception of mutual dependencies[303] in the relationship among players and between them and the resource, looking to the common future. The industrialists in the Liri Valley lacked such awareness. They did not conceive of themselves as a community but as profit-maximising individuals and acted accordingly.

In the Liri Valley, the entrepreneurs' perception of water property contradicted the very nature of the river. Flow, movement and interdependency are the key terms of river ecology: without movement and interrelationships with the bio-physical environment around it, the river would not exist. Waterpower itself is produced by the flow of water past (and inside) the mill-engines. It is an energy flow, not a stock. Obviously, the river could be fenced, and water stored in reservoirs, in order to ensure stable and measurable energy production. This kind of river appropriation, however, would lead to some kind of control over large tracts of the river by a single appropriator (a private corporation or State agency), a situation that theory defines as a *natural monopoly*. In this way, water would be allowed to flow only at the rate necessary for the production of power for the benefit of the same agents entitled to take advantage of it. Something similar, as Ted Steinberg showed in his *Nature Incorporated*, occurred in New England after the incorporation of the Boston Associates[304]. In that kind of system, not only privatisation, but also domination was required, both over the river and over other players, in order to produce waterpower and economic efficiency.

Neither the mill-owners in the Liri Valley, however, nor the Bourbon or the Savoy government, wanted domination. They sincerely believed in the Lockeian notion of economic freedom and national wealth and had left behind centuries of social domination in the form of feudal regimes. They imagined (and practised) a free access/individual appropriation framework, in which there was simply not enough space for the production of efficiency, both in economic and in ecological terms. Individual users tried to maximise their share of the energy yield by over-using the resource (obstructing the river-bed with stones and wooden fences, or installing engines within the river) at the expense of downstream use and that of the community. Modifications to the stream and to the river-bed, in a multi-owned resource, are extremely counterproductive, because they follow very narrow visions of the system and cause negative feedbacks to resource productivity and the players concerned. A classical prisoner dilemma occurs, therefore, because maximising individual choice means all the players lose.

Water disputes in the Liri Valley not only reflected local rivalries and power games; in fact, the river was the material terrain on which the transition to the liberal regime was being tried out and where social relationships left fluid by the end of the old regime were growing solid. This struggle for the control over water, however, was not a matter of class as much as it was of individuals and families. In this sense, water conflicts were indeed, as Neapolitan Law prescribed, a 'private' affair, where the 'public' (intended to mean the health and livelihood of people) was only instrumentally evoked by private contenders. Once the suit was settled, nobody cared any longer about either the river or the – defenceless – living world around it.

Finally, the Liri Valley case shows how river enclosures were part and parcel of the agrarian and made their own contribution to that general process of transformation of the society-nature relationship known as the Industrial Revolution. A fundamentally new kind of socio-ecological metabolism, the latter emerged in some regions of nineteenth century Europe, from there spreading throughout the continent and beyond, with unprecedented (and revolutionary) impact on the biosphere. Whether we term this new society–nature relationship 'industrial capitalism', or 'modern growth', or 'urban–industrial society', the fact remains that the enclosure of nature, its removal from the commons – as John Locke put it – was the starting point.

Chapter Five

Disciplining Water: Floods and Politics in the Apennines

And thus, since the long series of political calamities, every convenient order of physical and topographical circumstances in the country was upset.

C. Afan De Rivera, *Considerazioni* (1832)

This chapter will enter into the long-lasting tragedy of floods and *bonifica* in the Liri Valley, questioning the way environmental vulnerability was understood and dealt with by contemporaries.

During the nineteenth century, the perception of socio-environmental costs of land and water privatisation in the Kingdom of Naples mostly took the form of official reports on the part of civil servants. These reports were dominated by one single issue: floods. Violent inundations of the plain of Sora are recorded by the Bridges and Roads Bureau of the Two Sicilies in 1825[305], 1833[306], 1856 and 1857[307], while complaints about ordinary floods recurring in the rainy seasons are scattered throughout the entire period. With approximately the same classification criteria, the post-unification Italian State recorded major flood events in the years 1879[308], 1903[309], 1906[310] and 1910[311]. This chapter will try to understand how this long series of floods intersected with the political economy of both the Bourbon and the Savoy State. It will look at the 'disorder of water' with the rationalising eye of the civil engineer and will try to make sense of the unruly resistance repeatedly met with by different schemes for improving the river throughout the century.

To begin with, the very practice of classification, separating flood records from those concerning the 'economy of water', tells us something important about ideas informing the State's view of water politics. Though forming a whole socio-environmental reality, that of many rural economies and of people's lives in them, 'improvement' and 'habitation' had been separated into two conceptual realms, one referring to the benefits, the other to the costs of water use. 'Rural economy' had been translated into 'political economy'. This separation, however, was to be overcome through the supervising authority of the Minister of the Interior, the ultimate destination of the entire communication–administration flow. Thanks to the Minister's rationalising, all-knowing gaze over the provinces, the King – that is, the State – could exercise his benevolent compassion towards the country by means of new laws and/or *ad hoc* provisions. It was through this newly created

Disciplining Water: Floods and Politics in the Apennines

government infrastructure, that a new economy of water took shape in the early nineteenth century Liri Valley.

Generally speaking, schemes of land drainage and improvement fell under the category of *bonifica*. Among the different kinds of areas that formed the geography of *bonifica* in southern Italy, the Liri watershed pertained to that of the inland valleys, where the 'disorder of water' was perceived as a drama of permanent disaster-proneness. Due to their position upland, these valleys were also the areas of more intensive waterpower use. This kind of *bonifica* was much more controversial than the others[312]: it required a higher ability to control nature and maintain the artificial order through time, as more trouble could derive from an obsolete or ill-maintained reclamation scheme than from nature itself; and even in the absence of damage, an ill-retained scheme could cost money without producing results[313]. The economy of uphill *bonifica* was also difficult: it implied a much more complicated system of rate formation [*ratizzo*] between public and private owners and between upper and lower areas. Such a complex socio-environmental scenario invited to simplification. Not by chance, during the thirties the proposal was made of entirely abolishing what remained of the undivided commons [*demani*], by selling them to the agrarian landowners, in order to simplify tax collection and make a uniform rate system[314].

Despite Monticelli's paradigm remaining the most widely accredited explanation of hydrological risk, a different one was advanced by Neapolitan physician Salvatore De Renzi during the 1820s. This alternative paradigm, historian Costanza D'Elia has noted, emphasised the role of recent economic and social change throughout the country, and especially the intensification of water use, as the major cause of environmental damage[315]. One important difference between the two paradigms lies in where they draw political attention: to the uplands, where mountain communities and land parcelling were presumably devastating the country's forests; or to the lowlands, where capitalistic agriculture was intensifying land and water use, thus also increasing the social costs of Improvement.

The first recorded report of flooding in the Sora district – as we have seen in Chapter 2 – was that filed by the *Sottintendente* Giuseppe Massone in 1813. Breaking with the idea that hydrological 'disorder' had originated in southern Italy's ancient history and foreign invasions, the *Sottintendente* had clearly exposed local landowners and mill-owners as those responsible for recent inundations and invoked the 'arm of the Government' to force them to return the riverbed to its pristine state. Massone's report was only the first in a long series of similar reports over the following decades. Through time, the language, categories and concepts used to frame flood risk in the valley became somehow standardised and the civil servant's report became the preferred form of expression for a widespread interpretation of the political ecology of risk in the *Mezzogiorno*.

Seeing Like an Engineer

In 1825, shortly after being appointed as Director of the Bridges and Roads Bureau, the engineer Carlo Afan De Rivera was called upon to give a detailed report on floods in the Liri Valley and to suggest possible remedies. The town of Sora, he wrote, 'far from undertaking any expense to regulate waters and make them useful for turning machinery and for irrigation', tolerates the devastation caused by the 'unwise haste to grow cereals on steep land' and by the five mills situated on a short half-mile section, with the result that the difference in level between one and the next has been reduced so much as to make them all useless[316]. To reconcile public and private interests, De Rivera proposed to divert a side channel from a point well upstream, with a sufficient drop to be able to locate six mills there (five to be rebuilt for their prospective owners, one for the *Comune*, so that the latter had an interest in the affair); part of the same channel, upstream of the machinery, would be used to irrigate a large area of land, bringing in a huge income for the *Comune* of Sora. As regards the respective benefits, the expenses for one third of the flume and the whole irrigation channel would be met by the *Comune*, while the other two-thirds (as this was a *bonifica* project) would be incurred by the owners of neighbouring estates in proportion to their respective contributions of estate duty. 'It is of utmost importance', added the director, 'that the project be undertaken as soon as possible', with the *Comune* advancing the expenses for project design. The concern of the official as to the urgency of a solution was sadly confirmed by the flood of 11 December 1825, which inundated the entire town of Sora, providing, as one reads in a later report, 'a salutary warning that time should not be lost', due to the 'wretched speculation over preserving five mills', that could in no way offset the risk of 'seeing a town destroyed and many of its inhabitants dead'[317]. However, several months would elapse before Afan De Rivera's proposal, examined by the *Sottintendente*, the *Intendente* and the Council of State, held back by the usual bureaucratic machinations (this time attributable to the negligence of the courier), was submitted to the Minister, who would ask the Town Council to send its own deliberations to the mill owners to hear whether the latter raised any objections.

The Town's resolution expressed a long series of doubts about Afan De Rivera's project: besides the length of the channel (at least two miles to have the right slope), the Council was concerned by the need to build it in stone and supply it with bridges; a reasonable criticism to be levelled was that during the summer drought, with water flow being a third lower, it would barely suffice to power the mill wheels, while the river bed would stay dry, giving rise to stagnant water, which would make the air insalubrious, exacerbated by the municipal sewers that discharged into the same bed. The Council expressed a preference for the simple raising of the diversion dams to their original height, thereby eliminating the immediate cause of the rise in the channel bed and consequent water overflow. However, if the project were to be approved 'for better reasons', the councillors' observations would focus on splitting the expenditure. Indeed, the two thirds to be charged to neighbouring landowners should be charged to the mill-owners alone, as those who had created the damage. The *Comune* of Sora did not have an income sufficient to cover its needs and supplemented it with civic taxes charged to landowners: hence the third of expenses to be borne by the *Comune* would already be supplied by a new tax paid by the latter, who would thus end up paying almost the whole cost. The abuses committed by the millers with the raising of the level of the diversion dams would instead give the *Comune* 'the clearest right' to oblige them to lower them, which would make their millstones useless, since the stone columns that marked the original level were already below the current bed: the dredging operation would thus need to be repeated every year, at great expense. As things stood, the mills did not have sufficiently powerful water flows, and produced very little flour, which forced the population to procure flour from the mills of San Domenico and Carnello farther away. Moreover, the millers spent huge sums every year rebuilding the diversion dams destroyed by the floods and, after a few years, admitted the same Council, the situation would reach crisis point. All this therefore justified the greater interest, besides the greater responsibility, of the millers in the planned works, a situation in which the administrative intervention of the *Comune* would be decisive: it could force the millers to lower the dams, making them useless. To compensate them for the greater expenditure, however, the Council proposed that one third of the income obtainable from the irrigation channel rental should go to the *Comune* and the remaining two thirds to the mill-owners[318].

The whole question of water degradation thus seemed to hinge on the mills of the town of Sora. At this stage, it is worth seeing who the incriminated mills actually belonged to. Called upon by the Ministry of the Interior, the owners did not hesitate to make themselves heard: the first and most combative was the Bishop of Sora, who defended the interests of ecclesiastical assets, being supplied by a mill with an estimated income of more than 300 ducats. Describing the expense required by the project in question (which had not yet in fact been calculated), as such 'that it would upset, so to speak, the very finances of the government'[319], he proposed instead the simple reconstruction of the weirs to their original levels, a work impeded by the ban in the 1811 forestry law on woodcutting in the sur-rounding woods, against which the bishop was waging his own war. He concluded with the threat that should there be an 'abuse of power' by which the existing mills were destroyed, he would seek compensation for loss of assets. A copy of the charge was sent to the Minister for Ecclesiastical Affairs.

The curate to the monastery of Santa Restituta also deemed the project impossible to implement, given the paucity of the *Comune*'s financial resources. He further sought to show that his mill should not contribute to expenses insofar as the weir maintained the original level[320]. Once again the bishop, responding to the Mayor of Sora, praised the Council for the wisdom shown in rejecting De Rivera's project and added, not without some reason, that the real responsibility for raising the bed of the Liri belonged to the owners of land in the Roveto valley, through which the river flowed upstream of Sora, who grew crops on steeply sloping soils, generally weakening soil stability and leading to gully erosion. Finally, of interest is the reason given by the bishop to justify the raising of the dams: 'The silence of the council representatives, under whose eyes the raising occurred, authorised the operation'[321]. However, now the mill owners would be forced to contribute to an expense 'which, had it not been for municipal ineptitude, would have been far lower both for them and the public purse'[322].

The third mill belonged to the monastery of the Poor Clares, whose Superior complained of the frequent budget deficits and stated they could not be forced to contribute to expenses insofar as the monastery had been in possession of water rights since time immemorial. The responsibility for raising the height of the channel bed lay not only with the landowners in the Roveto valley, but also with the citizens of Sora who were accustomed to throwing building rubble into the river. Hence 'what justice can make an ancient mill owner incur the exorbitant expense of dredging the bed itself?'[323] Moreover, the citizens of Sora enjoyed the convenience of grain milling, while the owners earned little due to a huge estate tax. Finally, the fourth and fifth mills belonged to a private owner, Savino Marsella, who greatly praised the project, but did not join those interested in contributing, since his diversion dam was not located where it would be detrimental to the *Comune*, while, he remarked, 'there are many other causes behind the floods'[324].

The following June the Ministry forwarded all the documentation to the Bureau of Bridges and Roads (which had by then assumed the additional remit of Waters, Forests and Hunting): the response of Afan De Rivera stressed the reasons for the original project, pointing out, in response to the concerns of councillors, that 'everywhere one sees such canals made by private people to earn income with which to operate their hydraulic machines'[325]. As regarded the fear of stagnant water and summer drying of the river section running through the town, the director replied that the mills would be built outside the town so as to restore water to its natural bed upstream of the town. In any case, if the undertaking were not judged worth the expense, the dams for the mills would be forbidden. Strongly opposed by the administration and the local community, De Rivera aimed to redraw the order of things in the area and proposed to intervene in the river course with permanent hydraulic engineering works of a considerable technical level. Of course, De Rivera was first and foremost an engineer and set the greatest faith in man's capacity to manage and 'improve' nature. The environmental impact of the technology did not scare him; nor did his, albeit broad, view of the land include any perception of the chain of interconnections that would occur within the hydrographic basin as a result of channel straightening and water regimentation.

However, De Rivera's ideas had little likelihood of being carried through in the Liri Valley: aside from the chronic scarcity of funds in Bourbon and municipal coffers in particular, resistance to the project came not so much from the force of the establishment (religious orders), against which the government seemed intent on acting, but from the town councillors, whose main concern was, in this case too, the burden of the contribution to be incurred by the borough and, indirectly, by landowners. Indeed, it seemed that local society, with the landowning and taxpaying class as its spokespeople, was willing to accept the damage of water degradation rather than bear the costs of a radical technical solution whose spirit they did not share.

In the end, the Liri improvement project would be drawn up, after repeated requests following the floods of January 1827, by the Inspector General, Engineer Grassi: completely different from the project outlined by De Rivera, in an attempt to meet the municipality's need to contain expenditure, not even Grassi's project would find favour and the whole operation foundered. The inspector proposed the demolition of all the existing dams, leaving only one municipally owned mill and compensating the owners. A second mill was to be built after the river water level had been restored to its original state, downstream of that of the *Comune*, being fed by a small flume which channelled the overflow. Other mills could only be built on the River Fibreno. Although the technical implications and relative expenditure of this project were decidedly lower than those of De Rivera's, the town council opposed it: the machinery to be built on the Fibreno, objected the councillors, could not even be thought of if two bridges were not built first; with regard to the two mills on the Liri, everything was to be put off until Lake Fucino had been reclaimed, which might lead to the river channel becoming a marsh, in which case any dam would be damaging. Any decision was thus postponed until conditions beyond the sphere of competence of the municipal authorities were created[326].

The view of the *Sottintendente* was different: he was decidedly in favour of demolishing the mills. It emerged that none of the owners actually possessed either the waters or ground on which the mills were built, with everything proving to be an abuse favoured by past administrators[327]. He added that the Liri was not suited to irrigation and that the banks and palisades were placed by the neighbouring landowners not in the channel but on their respective land, for defensive purposes; and that traps used for fishing would be prohibited administratively with estate owners thus being obliged to clear the channel of trees, bushes and wood bundles obstructing the current. Here we have a further point of view, which could be defined as that of legality: those breaking laws and regulations, resulting in a major part of the environmental damage, were to be made directly responsible and charged with removing the causes of the water problem; improper uses of the bed (fishing traps) were to be abolished by law. The solution to the problem came from good laws and compliance with them in the region, not, as the *Sottintendente* seemed to imply, from painstaking cooperation between local actors.

Finally, all the documentation was sent from the Crown to the General Assembly of the kingdom for a definitive opinion[328]. The assembly rejected not only the Grassi project, but also the approval decided by the Town and the District Councils, and stated that De Rivera's original project was to be carried out to the letter, splitting expenditure according to responsibilities and respective benefits, which a commission appointed on the matter would resolve with rumination and 'without haste'. In exchange, the mill-owners would be given the new mills to build, the *Comune* would receive income from irrigation and freedom from flooding, better communication would ensue thanks to the bridges to be built and there would be improvements in climate, besides conserving the 'precious convenience of the nearby mill'. Owners of farmed estates would have soil fertility guaranteed for life, the flood risk being reduced. Millers would be freed from the expense of weir maintenance; the Fucino project would find the Liri river-bed ready to receive the copious waters from the lake. The landowners in the Roveto Valley, despite not receiving any apparent benefit, would nevertheless be burdened with the expense as a penalty for illegal deforestation and the ploughing of soils on the valley slopes[329]. The State Committee thus fully backed the views of the Director of Bridges and Roads, with a surprisingly clear-cut stance: it was, however, a statement of principle, not followed by initiatives.

De Rivera's project for the channelling of the Liri must be read against the grain of his broader political economy and ecology vision. As the clergyman Pistilli had done roughly thirty years before, the engineer De Rivera envisioned the Liri Valley as a place for experimentation with new ideas about nature–society relationships in southern Italy. In fact, the Liri project anticipated De Rivera's broader vision for the improvement of the country's nature, as he later described it in the chapter devoted to the Liri–Garigliano basin in his *Considerazioni*. Garigliano is the name that the Liri takes after the confluence of the Gari and indicates the low valley and delta of the river on the Tyrrhenian coast. De Rivera describes this as a highly fertile, but ill-drained and malarial area, where irrigation could be easily practised. Going up the rivercourse to enter the mid-Liri Valley, De Rivera's engineering mind became excited by the abundant resources offered by nature and just waiting to be properly valued: majestic mountains framed the valley, that could provide abundant timber were they not almost entirely tilled or devastated by goats; the plain of Sora could produce cash crops were it irrigated through canals from the Liri and the Fibreno; in Isola many more factories could be established by creating flumes to better distribute waterpower. All this improvement was prevented by maintaining the five grain-mills immediately upstream from Sora, which obstructed the riverbed and made the water level rise, subjecting the two towns to a permanent state of near-disaster[330].

De Rivera's narrative of the 'disorder of water' in the Liri Valley is one of political ecology, where the interest of the few (the five mills of Sora) stands against the general interest (improvement and habitation alike). However, while in Galanti's case the answer to water's disorders was sought in anti-feudal politics and property rights, when De Rivera wrote his book the latter had been long introduced. The answer was now to be sought in hydraulic engineering. Engineer De Rivera's gaze over the physical conditions of the kingdom is thus positive and forward-looking: restoring the 'natural' wealth of the country only requires injecting order into the 'disorder of water' and this can be done in two steps – by re-engineering the torrents into one-bed channels and by reforesting the Apennines. However, like his French colleagues at the *Eau et Forêts*, De Rivera's response to the issue of floods was largely based on blaming the poor for deforesting the mountains[331], especially mountain shepherds with their reputedly devastating attitude towards the forest commons.

One important difference between the projects of Pistilli and De Rivera is that only the latter's could be labelled as a *bonifica* – that is, a comprehensive scheme for both preventing the recurrent inundation of the Sora plain and for improving the use of water for irrigation and energy. Pistilli had only advocated fuller exploitation of the energy potential of the river, for navigation and waterpower. Thirty years later, the second part of his project had been largely carried out – albeit in a different fashion from that originally envisaged. Factories and mills were lined up all along the riverbanks such that no drop of water fell through the valley without having moved some piece of machinery. As an unforeseen consequence, this industrial transformation of the Liri Valley had contributed to the increased frequency and destructiveness of floods in the Sora district. De Rivera's project in fact originated from the need to solve this very problem by redesigning the structure of both water and water uses in the local space.

More than anything, De Rivera's was an engineering vision: he was inspired not only by the *need for*, but also by the *possibility of*, improvement through techno-scientific rationality as embodied by the civil engineer. He was heir to a long and glorious Italian tradition of hydraulic science and architecture, starting with Leonardo da Vinci and then Galileo Galilei, the undisputed father of modern hydraulics in Renaissance Tuscany. The nineteenth century hydraulic engineer was able to draw on long historical experience with hydraulic architecture in Tuscany, Lombardy, the Po plain and Venice[332]. Knowledge of such experiences and theoretical insights on the movement of fluids were available in books used at the royal schools of architecture and engineering[333]. The most important lesson that hydraulics experts of the nineteenth century could draw from their illustrious predecessors was twofold: on the one hand, Galileo and his disciples had emphasised the precariousness of any hydraulic architecture and the limits of mechanics in making the flow of water calculable; on the other, they had advocated what today's water policy theorists and makers call an 'integrated approach', that is a multi-disciplinary and multi-level view of water management in a watershed dimension[334].

Seeing Like an Engineer

At the time of De Rivera's appointment at the Corps of Bridges and Roads, Italian hydraulics was in a phase of transition. Increasingly the State aimed at improvements, especially as demographic growth and the commercial decline of the peninsula made it ever more imperative for new land to be drained and made cultivable. The example of other European states, especially Prussia, successfully pursuing aggressive reclamation schemes, stood before the eyes of Italian governors in the same fashion as the English enclosures movement, i.e. as signifying the inexorable path to progress. Hydraulics was not only, as the founding fathers had taught, a science for the defence of human habitation against water, but also one for the positive conquest of nature[335]. This political economy transition involved scientists as well; they began debating the need for a mathematical and theoretical approach to hydraulics, which responded to the increased demand for calculus and engineering. Such a theoretical approach could not properly account for the many variables (human and non-human) interacting with the flow of water in a real river basin; nevertheless, it promised to be a valuable tool in water projects. The debate went on for decades and interacted with changes in technology and the economy, such as the hydraulic turbine, urban sanitation and steam-powered navigation – all increasing the demand to re-engineer Italian watercourses in the second half of the nineteenth century[336].

De Rivera's plan to redesign a section of the Liri in order to prevent flooding, while also improving the use of waterpower, stood at the intersection of these many driving forces. The scheme emphasised the potential of technology in redesigning nature, without much concern for possible side-effects or unexpected consequences of this redesigned river, nor for its impact on the local landscape. At the same time, the plan had the ambition of regulating forms of access to and use of water, imposing a new rationality on and supervision over local life. The rationality of such a project was that of 'public interest', intended as the reconciliation of 'habitation' with 'improvement', since the latter had manifested a tendency to disrupt and destroy as long as it created wealth. The mind of the civil engineer was needed in order to balance creativity and destructiveness in the local space.

But this rationalising mind did not enjoy social consensus. Although what was lying before the eyes of the civil engineer was hardly pristine nature, not even in the case of peripheral inland areas like the Liri Valley, De Rivera's channelling project was perceived as a disruptive novelty. Of course, the mill-owners opposed it, for it would have interrupted their long-standing possession of water, forcing social costs upon them; but there was certainly more than this in the opposition of the Town Council and thus of the majority of local landowners. One consideration, completely absent from De Rivera's view, seems crucial: the risk that channelling one river section would increase the likelihood of floods downstream. De Rivera quickly rejected this criticism, stating that 'in many places you see this sort of canal built to make waterpower available to mills' – thus completely glossing over the question of downstream effects. The feared increase in flood risk due to channelling was not so exaggerated: approximately twenty years later, Engineer Elia Lombardini, a major Italian hydraulic scientist of the age, wrote extensively on the issue in the pages of the journal *Il politecnico* and his observations later became science through his widely adopted textbook *Idrologia fluviale ed idraulica pratica* ['River Hydrology and Practical Hydraulics], published in the early 1870s[337].

Uncertainty and precaution were not part of De Rivera's vision of how water degradation could be repaired in the Sora district, a vision that was entirely positive and acknowledged no apparent cost; and yet they were foreseeable and real to local people and could be employed as convincing arguments against the re-engineering of local space. A disorderly but 'natural' river was perceived as better than an artificially ordered one; though the Liri was no longer a 'natural' river, the ideology of the pristine could be successfully employed to defend the status quo, especially as local powers defended their exclusive control over the local landscape. The Head of the Water and Forest Bureau, a gigantic bureaucracy potentially interfering with the entire rural economy, De Rivera was certainly seen as a threat to both local powers and to the moral economy of the peasants. The decades following the fall of the French empire were a period in which forest laws and bureaucracy came under attack by both liberal politicians and local administrations, who exposed corruption and the inability to take local needs into account. 'Let us get rid of those institutions that have been found harmful and useless', deputy Pascucci exhorted the Neapolitan Parliament in 1820, 'first among which is the forest administration. It has destroyed the forest and robbed the people'[338]. De Rivera, in turn, insisted on the need to strengthen the Administration of Water and Forests by forming a highly reliable technical corps of keepers and incorruptible civil engineers, in order to have the good laws enforced and respected.

How could the rationalising mind of the civil engineer find such insurmountable barriers in the local sphere? Unlike the case of the Rhine channel straightening, it was clearly the absence of a strongly imposing will on the part of the State that permitted local resistance against environmental engineering to win and the river to remain un-channelled and even un-embanked for a long time to come. It is now to this un-improving State that we shall turn our attention in order to make sense of how floods intersected with political economy in the nineteenth century Mediterranean.

The Un-Improving State

At the time of Massone's first report on the Liri Valley floods, the government that was invoked to return the river to its pristine condition was French – that of King Murat. Probably due to the pressure of either military emergency or budget restrictions, the Empire proved a far less improving agency than it had announced itself to be. Despite its importance as administrative capital of a frontier district, Sora was unable to elicit due attention from the State to its hydraulic problems, even for the mere reconstruction of the destroyed bridges. Indeed, the related expenditure was split between the government, the *Comune* and landowners, but the government funds, allocated by the February 1808 decree, had not yet been supplied[339]. Under solicitation from the Director of the Bureau of Bridges and Roads, the government allocated a further sum for *bonifica*, as an advance payment for expenses incurred by landowners. The sum would be repaid in instalments through the '*ratizzo*'[340] system, but this time too, perhaps due to a technical glitch, the funds never arrived and the whole matter came to nothing.

Nevertheless, inadequacy in responding to pleas for help against flood risk by the local society was not a feature solely of the colonial State. The whole Bourbon Restoration period is equally interspersed with a similar pattern of disasters and lack of State intervention. This was not, as one might think, for lack of agreement as to the need for the State to oversee the redressing of environmental risk; rather, it was the State's non-compliance with what was unanimously reputed its most sacred function, that of preserving the life and integrity of its citizens. The story of floods in the Liri Valley is also one of Bourbon failure to fulfil the myth of the King's thaumaturgy[341].

In 1833, eight years after De Rivera had left the Liri Valley, another major flood hit Sora. Describing it as 'particularly violent', the district council brought up the whole question at the provincial council, as if years of debate on the same subject were worth nothing: the provincial council could do nothing but note that it was a question of an administrative controversy, concerning damage caused by mill owners (but with no mention being made of deforestation), and in turn the Minister for the Interior looked to the King to take 'appropriate measures'[342]. But by this point it had become pure administrative rhetoric: initiatives for *bonifica* in the district continued to be postponed, in a sort of bureaucratic ritual consisting of a continuous bouncing-back of opinions, responsibilities and decisions between central and peripheral bodies. Ten years later, in 1842, the district council was to complain of 'the unceasing and growing devastation' caused by mountain deforestation, 'the deleterious effects of non-compliance with relevant laws and regulations'[343]. The provincial council of *Terra di Lavoro* was unable to solve such a complex question, which would most of all require coordination between upland and lowland economies and agreed regulation of the freedoms that each landowner deemed sacred. As in an extenuating political ritual, which fails to become real local government practice, each year district and provincial councils continued to re-propose the question of '*bonificas* demanded' in the district of Sora, without ever making precise proposals, leaving the matter to the government.

At the beginning of the 1850s, however, the situation seemed to be reaching a turning-point. It was the municipality of Sora that directly took on the responsibility for a project to defend the town from the floods on the Liri by means of a series of embankments. The novelty lay in the fact that the council stated it was willing to bear much of the expense, thanks to an advance payment made by a councillor for 4,000 ducats (at five per cent interest) out of a total of 15,000 ducats. It should be pointed out that the council of Sora had an annual spending capacity of about 1000 ducats. The new *Sottintendente* was greatly in favour of the project, proposing the institution of a new duty for years to come to pay off the debt. For the first time, local government institutions appeared set on taking initiatives, although they might damage the more or less entrenched interests on water use in the district. What made it urgent was undoubtedly the gravity of the disasters the town experienced with the arrival of each autumn. As the *Sottintendente* explained in April 1852: 'I found both a colossal and necessary task pending here: to ignore it further would be seriously injurious for this capital town which suffers so many tragedies due to the frequent floods', caused by the materials transported by the Liri from the surrounding mountains and deposited at Sora due to the frequent river bends in the valley, the narrows and the height differences 'which even make the waters rise in the middle of the town'. The problem had been long outstanding, continued the *Sottintendente*, and repeatedly buffeted by financial problems. Thus he intended to seriously take up the matter 'that makes the inhabitants live permanently in dire consternation'[344].

Despite the auspicious start, this project also failed to take off. The stumbling-block on this occasion may well also have been financial, possibly the lack of a loan. As is stated in an 1856 document, the *Comune* 'was not in a position to contribute to the above-mentioned embankment scheme'[345]. A subsequent general project for land reclamation in the Liri river basin, commissioned by the Ministry of Public Works, was likewise blocked due to lack of funds[346], and nor did the Provincial Deputation for Public Works find the sum of 1000 ducats to pay for project execution in advance[347]. After a severe flood in November 1857, the King intervened in the question, ordering the province to advance at least half the sum 'from any fund', given that 'because of recent disasters caused by that river in the town of Sora the authorities of this province are demanding the work of embankments', while the General Administration of *bonifica* pointed out that 'without the wherewithal, nothing can be done'[348]. This was towards the end of a reign on the verge of swift collapse, and the emptiness of the statements of intent regarding *bonifica*, continually contradicted by scarce endowment – or a total absence – of funds, came as no surprise.

The final years of the kingdom of the Two Sicilies were also the most tragic for the environmental situation in the Liri Valley. Floods were recorded in December 1856 and November 1857 when the council of Sora, in a plea to the Ministry of the Interior, denounced the fact that bankfull discharge on the Liri, which had occurred in the past on the occasion of extraordinary rains, had become frequent even after normal autumn rains, due to the progressive filling of the channel, whose level had become higher than that of the town. Since the beginning of the season, four floods had already hit Sora, affecting a population of 12,000[349]. The most terrible, however, was that of October 1858, which drove *Sottintendente* Colucci to turn once again to the *Intendente*:

> Sir, with great regret I take up my pen to point out to you a matter which has always been of the utmost concern to me and which, however far back it goes, is also injurious to the population of this *Comune*, which is capital of an interesting district. The heavy rains of the 13th and 14th days of this month brought home, yesterday and today, the painful yet recurrent spectacle of the floodwaters of the Liri invading Sora throughout its length and breadth, completely blocking off all routes for the space of twenty hours. The fear and the damage afflicting this unfortunate population make my words redundant. In this respect I have written and done everything that was in the limits of my competence and perhaps even more...[350]

The *Intendente* took up pen and paper and wrote to the Minister for the Interior, the Police Chief, the Director General of Land Reclamation and the Minister for Public Works. While the letters were travelling between Sora, Caserta and Naples, a new flood occurred between 27 and 28 November, that was 'so impetuous and imposing that none of the inhabitants remembered the like'[351]. Indeed, for the first time, the floodwaters, besides entering the town from the sides, submerged it from the upper part, 'such that the river, as if the town's streets were its bed, ran with such violence and devastation that it filled even the most courageous with terror'. At various points the height of the waters exceeded eight palms, damaging the workshops on the lower floors of buildings. Then, with the bridge to Naples broken, the waters flooded the road, interrupting the postal service; and inundated neighbouring land, leading to arable crop loss. As the waters withdrew, they left in the streets of Sora about four palms of mud. To remove this, required waiting until the end of the rains, 'with immense detriment to public health'. In the meantime, the river remained 'swollen and impetuous', preventing repairs to the bridge and the road for days to come[352].

The requests for, and promises of, help continued apace in the pages of provincial correspondence. In the never-ending saga of flood security in Sora, the Ministry of Finance was involved after requests made by the Ministry of Public Works, given the chronic shortage of funds of the *Bonifica* Bureau[353]. However, the exchange of papers that accompanied every disaster was absolutely futile as there was insufficient financial cover to pay for the works (estimated by the *Bonifica* Bureau at 16,000 ducats[354]). The *Comune* of Sora stated that it could allocate 4,000 ducats, in four annual instalments of 1000, but only starting in 1861: so what had happened to the funds which the Council said in 1853 it was ready to advance for embanking the Liri? It emerged that the *Comune* had decided to contract a debt for a much greater sum (10,000 ducats) to purchase 'a house for the use of the Jesuits' and that the Council's budget was thus committed to repaying the debt in annual instalments of 1,500 ducats, plus five per cent interest, until 1861. To meet the commitment, the Council had already raised all consumer taxes and jurisdictional proceeds.

Thus, to sum up, faced with a situation of chronic river degradation, the town of Sora decided to invest a sum equivalent to two thirds of the whole cost of the improvement scheme for a different allocation, committing itself to a burdensome financial effort for almost a decade to come. The strategy was clearly to ensure that payment for embanking works be levied from provincial and state funds, which did actually occur, albeit only after a decade of disaster and destruction, and at the cost of sacrificing a broad approach to solving the problem, instead only carrying out works that were strictly necessary and urgent. The provincial council finally decided to levy a surcharge on the district estates to allocate to *bonifica*, but the funds thereby obtained only amounted to 2,400 ducats. As lamented by the Commissioner for the Terra di Lavoro, although the Royal Decree on 11 May 1855 established the splitting of expenses between all parties (*Bonifica* Bureau, *comuni*, landowners, province and the general treasury), 'it is inevitably observed that in almost all works carried out in this province, the necessary expenses are incurred only by the provincial authority, or at least for the most part'[355]. A further despatch of correspondence then ensued between the Commissioner, the *Bonifica* Bureau administration and the Ministry of Public Works to establish who should contribute to *bonifica* and with how much. However, it is inappropriate to speak of *bonifica*, since the works executed in the end (in 1859) consisted only of modest embankment works along the walls of the town and repairs to one of the bridges.

Despite being an area of frequently recurring flood events, the Liri watershed never had a comprehensive reclamation scheme, aimed at regulating water and forest use. The whole matter of periodic and recurrent flooding was subject to crisis management and affected by the reluctance of the various competent authorities to allocate funds, partly dictated by the problem of protecting vested interests. This was how newly appointed Prefect Homodei, representing the King of Italy soon after political unification, interpreted the situation in Sora. An engineer with previous experience in the Water Bureau of the Valtellina (in the province of Pavia, northeastern Italy), Homodei was struck by the fact that no final reclamation scheme had ever been compiled for the district[356].

In the years surrounding Italy's Unification the blockage concerning the matter of embankments was freed: a law on public works passed in 1865 assigned embankment constructions to the State, through the provincial units of the Royal Corps of Civil Engineers [*Genio Civile*][357]. This gave a swift impetus to construction of the Liri river embankment between Isola and Sora, the work being farmed out to local contractors.

Nonetheless, the flooding continued, partly because in the meantime the Liri had started to collect outflow waters from Lake Fucino. The causes of river degradation continued to lie upstream of all discussions and projects and remained, as denounced by the Chief of the Royal Corps of Civil Engineers of Caserta, the 'ill-advised crop farming of overlying sloping land, a regrettably frequent cause of serious consequences in southern Italian water regimes'[358]. In 1862 the *Comune* of Sora again allocated a sum of 60,000 Italian lire for works deemed most urgent (demolition of masonry bridges to be replaced by iron bridges, straightening part of the river bed) and the State seemed to be willing to contribute as well with a special grant of 16,000 lire. However, the Ministry of Public Works wanted a consortium to be established to tackle the roots of the *bonifica* problem throughout the district and not only with temporary measures. The idea was roundly rejected by the Deputy Prefect and the *Comune* since it was perceived as Utopian and liable to bring about serious delays.

Although the matter of a consortium was not raised for another decade, the Prefecture of L'Aquila, which was also responsible for the works at the Fucino outflow, further slowed down the works: engineers in Sora repeatedly requested the blocking of the discharge into the Liri of fifteen cubic metres a second, from the lake which was being reclaimed, in order to carry out embankment works. The matter was far from simple, however. Papers once again danced between the two Prefectures (L'Aquila and Caserta) and the respective offices of the Corps of Civil Engineers, the Ministries of Public Works, Public Finance and Agriculture, Industry and Commerce, the latter an organisation belonging to the ex-*Bonifica* Commission, charged with overseeing the transition of the Fucino–Liri question from the Bourbon to the Italian government. While the correspondence did its rounds, the waters of the Fucino continued to flow into the Liri's channel, aggravating an already compromised situation of hydrological equilibrium downstream. Architects argued about the problem of measuring river flow: this was important to deciding whether or not the discharge from the reclaimed Lake Fucino compromised the water regime in the Liri. However, the change was not merely quantitative: it was observed that material from diggings upstream, deposited by the bank on a bend in the river, were being carried by the current during bankfull flow and blocking up the lower section, with more meanders, greater changes in altitude, hydraulic works, barrages and side-channels. While this could have been the real cause of the Sora floods, the question was entrusted to an aged technician, and never examined properly. The discussion then returned to the question of downstream compensation, rather than the causes upstream. In this case too, initiative was slow to follow the paper trail. After more than two years of bureaucratic procedures – and the Prefect's concern about the threat to public order from the agitation of the population (which in truth limited itself to sending a petition to the Ministry of Public Works), and about the four telegrams a day he received from the Mayor of Sora – it would be discovered that the slowdown was due not only to conflicts of competence and bureaucratic elephantiasis. Much more concrete opposition came from the director of the land reclamation works being carried out by Casa Torlonia, with a considerable capital investment, in the valley of Avezzano: things had been done without consultation and the expenses would once again be paid by the wretched population of Sora. As soon as they had begun, in September 1865, the embankment works were again suspended and ruined by the current since no agreement had been reached with those commissioned to drain Fucino.

But that is not the end of the story of *bonifica* projects in the Liri Valley: as had already happened in the past, the question lay dormant for several months, perhaps by virtue of some temporarily effective expedient, to then rise up from the ashes the moment it was decided, on this occasion by the newly-formed Provincial Deputation of Terra di Lavoro, in December 1866, to look at the studies of the Liri once again; the authorities concerned were asked to supply the relevant documentation. The story thus continued in this vein: in 1869 a consortium was established between the deputation and the *Comune* of Sora; a government subsidy was allocated on the basis of the Law on Public Works; a new embankment project was set up by the Royal Engineering Corps of Caserta in 1875, also obtaining a decree of statement of public utility; but once again the allocation was lost among bureaucratic delays and mishaps. A contractor abandoned the works and demanded that his deposit be returned, a new call for tenders was deserted (this was already two years on, in 1877) and in 1879 again serious floods led to the Deputation expressing 'the gravest concerns', which called for a new project to be set up[359]. On 16 July of that year 'the instant rise of the Liri river waters due to the discharge of the Fucino waters [...] caused at Sora flooding of sixty centimetres, which submerged all the provisional works being executed'[360] and the deputation asked the Prefect to put pressure on the Minister for Public Works and the Prefecture of L'Aquila to 'intercede with the Fucino concessionaire' to regulate discharges so as not to cause further damage. For his part, the director of the Fucino Opera continued to deny that there was any direct relation between the now completed reclamation works and the Sora floods: calling attention to the concession contract and the criteria established therein, the representative of Prince Torlonia opposed the real river with a flow of legal arguments, invoking the 'naturalness' of the works performed, Lake Fucino being a 'natural tributary'[361] of the Liri. In the meantime the Provincial Deputation paid a total of over 312,000 lire for the failed embankment of the Liri, the *Comune* of Sora paid annual instalments of 15,000 lire and, in 1882, the State decreed the allocation of a further subsidy of 25,000 lire.

In that same year, the Jacini enquiry arrived in the district of Sora. Local surveyor M. Mancini reported a population of roughly 150,000, whose growth in the last decade had been hampered by famine and epidemics and especially by recurrent outbreaks of malaria, killing hundreds of people at once. The problem was particularly acute in the lower part of the valley, downstream of Isola, where malaria was linked to the 'deplorable conditions of the watercourses', regularly overflowing their banks during the rainy season. It was clear enough to the writer that,

> until well-ordered, effective drainage and reclamation works have been completed; until a more rational regime for irrigation has been established by law; until timber felling has been regulated according to higher principles than those adopted by private owners and local communities, a terrible threat will always be pending over our rural population[362].

What Mancini advocated was the firm intervention of the State both upstream and downstream, as the only possible means to establish order among the conflicting interests of non-regulated 'improvement' and 'habitation'. With private property firmly established in the Liri Valley, experience had shown that, contrary to what enlightened reformers had predicted, private and public interest did not automatically converge.

The Disorder of Industry

Civil servants' reports from the Sora district painted a picture of the 'disorder of water' as a recurrent source of impending disaster, due to both lack of water use regulation and lack of State intervention. As for the remote causes of disaster, these were distributed among a variety of actors, from those upstream (the Fucino project and upland communities) and through the neighbouring areas (agriculture in the Roveto valley) to lowland irrigation in the Cassino plain. All things considered, the picture of environmental degradation had definitely become more variegated than in earlier times and deforestation was just one of the causes. Nevertheless, something still remained absent from the picture and this was industry. After De Rivera had unsuccessfully charged the five mills of Sora with causing floods, nobody had dared to blame them on the factories that overcrowded the mid-Liri Valley, with their disorderly network of mill-dams and mill-races. That is, nobody except the industrialists themselves. Surprisingly enough, it is from their own account that we hear another tale of disaster, one different from that found in civil servants' reports.

In this alternative story, disaster is but one affordable cost of the creation of wealth. Industrialists can occasionally have their properties inundated, as happened in the flood of 19, 20 and 21 January 1845, when the greatest damage was suffered by the Ciccodicola factory, at Le Forme; by Polsinelli, located inside the Ducal Palace of Isola; and by Courrier, whose paper mill used the same waterfall[363]. Nevertheless, experience showed that such costs were borne everywhere in water-powered factory systems such as that of the Liri Valley, since these operated in an inextricable symbiosis with water flow (as seen in Chapter 3) and could not be conceived at any distance from it. After De Rivera's idea of channelling the river and creating a network of artificial waterpower canals had been rejected, harnessing the Liri and Fibreno had remained a matter of building (and rebuilding) mill-dams and millraces, which were easily destroyed by the current, on an almost seasonal basis within the river channel. Keeping things disorderly had been a way for local industrialists to enjoy their freedom to use water as they pleased. No despotic hydraulic society had emerged in the valley, no centralised bureaucracy, but a rather chaotic concentration of factories, superimposing itself on the available space along the river.

Furthermore, no cooperation seemed possible among the appropriators, who appeared as if forced to fight one another all the time in order to ensure possession of their water. When, for example, in the midst of a period of appreciable increase in flood risk, the Town Council of Sora finally seemed willing to plan a reclamation scheme, it was the manager of a paper mill, a certain Engineer Bucci, who promptly submitted a project. The District Council asked the project to be examined by a government expert, with the task of establishing whether the works indicated could be carried out 'in a river so full of hydraulic works'[364]. In the meantime, however, the project began to encounter opposition, this time not from monastic orders or farmers, but from industrialists, namely the De Ciantis brothers, who had had a long dispute with the *Comune* and with other owners over their hydraulic works on the Liri. They pointed out that the Bucci project would destroy their woollen mill, depriving it of water; according to them, it was a futile scheme that would have the sole effect of making over 400 workers lose their jobs.

To industrialists in the valley, flood risk became only one aspect of their complicated interactions with one another in the local space. Examples of this abound. Look, for instance, at what happened to mill-owner Andrea Tuzj of Sora between 1870 and 1872. After building a factory 'producing wealth for the town', on an estate that he possessed along the River Liri, Tuzj was hit by a Prefectural decree, upon reports by 'some ill-wishers in the town', ordering the demolition of some works, in particular a 'provisional palisade structure' constructed to protect the factory from any floods[365]. Against the decree the industrialist invoked the Law on Public Works, which appeared to have been erroneously applied by the engineer of the Royal Corps sent to inspect the site by the Prefect, insofar as Art. 168, par. F, on which the decree was based, referred to embanked rivers, not to those embedded between high banks like that section of the Liri. Upon the appeal from the industrialist, the Minister for Public Works appeared to take the side of Tuzj, who invoked in his favour the damage arising from the suspension of work at his mill for seven months, the deterioration of the works (which were provisional) and the prospective miseries of the working class[366].

In his own defence, Tuzj supplied a fairly clear picture of the mindset of entrepreneurs and of the framework in which they operated:

> From experience we are taught that such works are permitted, when not detrimental to the public, or to private people, despite some slight transgression of the regulations in force, in view of the benefits to be gained. If this were not the case, what a large number of factories one would have to demolish! Setting up a new factory [...] is

not only in the private interest, but also in that of the State for the financial side of things and the encouragement of industry, and of the whole public.

After ascertaining that Tuzj's hydraulic works had already been declared damaging to the water regime by the Bourbon Division of Waters and Roads twenty years before, the Minister ordered the Prefect to verify whether the works had really been so damaging. In the end, he repealed the Prefect's decree, stating that the palisade of Tuzj was not only regular but also useful. It emerged only two days later that, besides the palisade in question, the *Sottintendente* of Sora had found 'three flow barriers extending into the stream and various heaps of rocks': it is not clear what Tuzj was actually building close to his mill, but above all not clear whether such works could alter the course of the river in the downstream section, so that 'an expert's opinion of this' was required[367]. The expert, Engineeer Bianchi from the Royal Corps, visited the site the following November and reported that, although all evidence, 'as far as might be known from public rumours', pointed to Tuzj for many breaches of the law, there were no precise depositions against him[368].

Moreover, in the course of the dispute, it had emerged that the Tuzj question was representative of the *modus operandi* of a whole entrepreneurial class. Accused of having dismantled the support for the third span of Corte Bridge, in his defence Tuzj stated: 'In time of floods it is consolidated practice to build a provisional dam at Corte Bridge so as to better channel the Liri waters to the left bank'[369]. In other words, his initiative came within the accepted practice by which users of the Liri defended themselves from changes to the river regime. At this point the minister wanted to know from the Prefect 'whether or not all the uses made of the waters were legitimate'. If not, he would take measures against illegal users. Here the matter founders on the chronic lack of information afflicting Prefectures' water management.

Something similar had also happened to the woollen mill owned by the De Ciantis brothers, which was confiscated following the firm intervention of the Mayor of Sora, due to unauthorised hydraulic works constructed along the course of the Liri. In their response to the measure, the De Ciantis complained that, after the floods of the previous year (1865), the Bureau of *Bonifica* had undertaken to raise the river embankments, building 'high palisades'. Due to this and the discharge of waters from Fucino, floods had destroyed the old embankments erected to protect their factory, forcing them to rebuild them higher than before. The De Ciantis had been immediately reported by their neighbour, Andrea Tuzj, and the Prefecture could not but order 'reduction to the original size', after consulting the Engineering Corps. The fact was that the De Ciantis had ignored the procedures established by law in such cases; in other words, they had not applied to the Ministry of Public Works to obtain the authorisation for defence works. In reality, they had acted according to common practice among industrialists of the Liri Valley, who were used to considering the river as directly part of their property and its use as 'free' and unconditional. The laws of the Italian state, however, offered fewer margins than before for behaviour of this type and required greater attention in respecting various protocols established by the Law on Public Works, passed in these same years. The De Ciantis thus appeared guilty of negligence in failing to take due heed of the administrative change under way and to take the appropriate steps in time[370].

In 1888 there was still talk of a definitive *bonifica* project for the Sora district, in the absence of which the provincial deputation refused to invite tenders for provisional works, like those requested by the *Comune* of Sora for the drainage of stagnant water in the village of Carnello[371]. This was an area with the highest concentration of factories in the Sora district, with many hydraulic channels, canals, islands, weirs and diversion dams. The hydraulic degradation of Carnello, resulting from intensive unregulated river use by entrepreneurs in an area already affected by river degradation, constituted a particularly delicate aspect of the *bonifica* question. Its solution, as we saw in the previous chapter, would be a long time in coming and would meet the tacit resistance of entrepreneurs, blindly attached to the concept of private water rights. Property rights, however, do not in themselves conserve the environmental equilibrium, prevent damage or ensure the reproduction of the resource on which they are exercised, especially if the resource in question is water. The difficulty in establishing forms of cooperation between landowners and local authorities (consortiums) is undoubtedly a prime constant in the history of hydraulic degradation in the Liri Valley.

In August 1900, for example, the question of the Fibreno was once again proposed by some factory-owners who occupied the left bank, including Enrico Polsinelli, and the Engineering Corps reported that the works requested did not fall under those classified by the Law of 18 June 1899, which was why all those concerned had to form a consortium[372]. Nothing more was heard of the affair until November 1906 when, after continued heavy rainfall starting the month before, the river overflowed into the village of Carnello, chiefly damaging fields and houses in the eastern area. Here the water reached two metres high and twelve days after the flood the soil was still waterlogged. East of Carnello, there was indeed a depression from which the waters found it hard to drain away, which was periodically affected by flooding and ponding. However, the cause was not only the topography of the place, but also the fact that defence works carried out in the past were completely inadequate: the river bed was subject to continuous raising due to the high density of dams and weirs and hence the Fibreno at that point flowed at the level of the surrounding country, flooding it when the minimum bankfull discharge occurred. It was a substantial change in the state of the river basin, arising from changes over several decades of industrial exploitation of the river, that had profoundly altered the relation between the river and the surrounding country. The Deputy Prefect of Sora, in informing the Prefect of the situation after the recent floods, added cautiously that 'as a contributory factor in the deplorable flooding was the channelling of water used by privately-owned factories into the Fibreno and changes in river banks or possible illegal works', a visit by the Engineering Corps would be useful, not least to avoid public disorder[373]. In this case, the embankment works would fall within third-category hydraulic works provided for under the Law of 25 July 1904 and it would thus be the Mayor's responsibility to set everything in motion.

After the customary bureaucratic delays, in 1910 the owners of land damaged by the umpteenth flood reported the Council's inertia to the Prefect. However, in reality this lack of initiative was tied to the difficulty of getting the local authority to agree with the landowners and the latter amongst themselves. The Mayor accused some of them of having tried to get the works executed in their own personal interests, attempting to slip them through as *bonifica* works. For this reason, it was not even possible to agree on the most urgent works to carry out. The despatch of yet another engineer from the Royal Corps was to prove futile, as he was highly unlikely to be able to unravel an almost century-long conflict in Carnello, while the *Comune* sought the intervention of the provincial deputation.

Five years further on, in August 1915, the *Comune* of Sora was assigned to a special State Commissioner: in a heroic attempt to impose some order on the matter, the Commissioner asked the Prefect for a list of flumes in Carnello. The answer came back that to conduct this research in the archives it would be necessary to have the names of each concessionaire. This was tantamount to saying that no list of this kind had ever been compiled and that the state apparatus was unaware of what happened on the Fibreno, both in legal terms and in reality. In the annual succession of floods, the Engineering Corps restated that a consortium had to be set up, or there was no prospect of finding a far-reaching solution coordinating works on the Fibreno with those on the Liri.

A consortium for embanking the River Liri had been founded as a Charity [*Opera Pia*] in February 1904. Apart from the paper manufacturer Emilio Boimond, no other entrepreneurs were associated with the consortium[374]. At the same time the *Comune* of Sora had put in an application to classify the channel straightening works upstream of the town in the third category and was waiting for a decree regarding this. The works chiefly concerned the village of La Selva, where the flooding of the Liri, which was very winding in this section, was a recurrent phenomenon (floods in December 1903 and November 1906 were particularly violent[375]): the village had probably experienced urbanisation in previous decades, which had increased the risk of hydraulic hazards.

Indeed, in the course of this history of river degradation, it was not only the river that experienced physical changes. The towns also changed and expanded. The population had increased and Sora and Isola had become poles of attraction to the neighbouring villagers[376]. Therefore the risk threshold shifted during the century. Indeed, as occurs in all alluvial areas with dense urbanisation, due to the scant consideration paid by poorly-regulated urban expansion to the phenomenon of river degradation, the risk had actually amplified. The industrial transformation taking place over the course of a century had thoroughly reconfigured the place as well as people's relationships with it. Despite its being continually invoked and referred to as an undisputed point of reference, no 'pristine' state of the river could ever be restored and people just had to adapt to the new socio-ecological order that industry had imposed over the valley.

There is no happy ending to this story of floods: to this day, the Liri–Fibreno watershed is included within the Special Plan for Hydraulic Risk of the Watershed Authority entrusted with planning risk-prevention measures in the interconnected basins of the Liri–Garigliano and Volturno rivers (figure 6)[377]. What did come to an end, at some point, was industry itself. Starting with a serious crisis just before the World War One, slowly but irreversibly the factories along the river began to close or move elsewhere. Those that remained installed electric engines and sometimes rearranged the old mills into small power plants. On the other hand, the propertied classes of Isola and Sora managed to save their watershed from complete

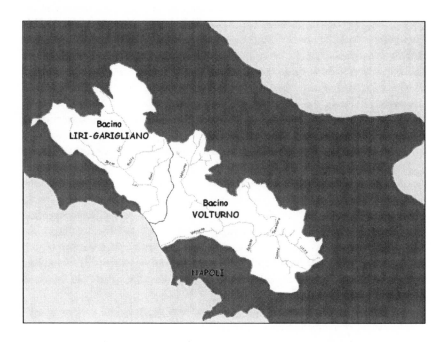

Figure 6. Map of the Liri-Garigliano watershed, 2006
From: Autorità di Bacino dei Fiumi Liri-Garigliano e Volturno, *Piano Stralcio per l'Assetto Idrogeologico, Rischio Idraulico: Bacino del Fiume Liri-Garigliano* (April 2006), http://www2.
autoritadibacino.it/.

re-engineering: no straightening or levelling of the bed was carried out and only minor embankments were constructed in the town centres. In the 1950s, attempts to build a big power plant just upstream of Sora (transforming the La Posta lake into a reservoir and depriving the Liri River of the Fibreno's inflow) were strenuously opposed and ultimately defeated, by means of erudite conferences, painting exhibitions and poetry, as well as street demonstrations and political action. With the Isola Liri waterfalls being designated as 'national heritage', the area won its democratic battle against big industry and the State; in addition, local industrialists maintained their energy autonomy. The 'natural' order of local things was safe[378].

***

This chapter has sought to make sense of what happened to the idea of 'disorder of water' in the course of the great transformation taking place in the Liri Valley with the advent of political economy. One important consequence of the age of Enlightenment in southern Italy – as we saw in the first two chapters – was the idea that floods were *not* acts of God: they were indeed a political issue. Initially, the idea was that the 'disorder of water' was the result of twenty or more centuries of foreign domination and political barbarism, in the dual form of feudal power and communal tenure systems. To restore nature's order required an act of political revolution – the overthrow of feudality. After the fall of the French Empire, though, the flood discourse began to change. The long advocated political economy was now undisputedly ruling the country, yet the 'disorder of water' had not ended. Upland communities were still there to be blamed, of course, but a more complex scenario of responsibilities was also emerging and it was a very slippery one. Private property, not its absence, could now be blamed for environmental damage and social costs. Private and public interest did not always converge. The arm of an improving State was badly needed to impose order on the environmental mess afflicting the country.

But the State was weak and financially broken and too many swamps were there to be reclaimed. In the end, those ruling southern Italy in the eighteenth and nineteenth centuries were un-improving States, in the sense that their schemes for redesigning the country's nature–society relationships were only scantily implemented; nor did they receive high priority among the State's economic policies. In a sense, the story told in this chapter seems to be a mirror image of the 'conquest of nature' being waged in other areas of Italy and Europe from the early modern era onwards. Unlike Venice and the Po Plain, Holland and Denmark, the Oderbruch or the Rhine river basin, most of southern Italy was sloping rather than flat; its story may be considered a Mediterranean version of reclamation: that is to say, one where nature wins and the State gives up on its improving schemes. Indeed nature plays an important role in this story: it is the Apennine ecology after all, with its ever-rising mountains, its volcanic soil and its unpredictable, torrential rainfall, that largely accounts for the failure of human attempts to master the environment.

Disciplining Water: Floods and Politics in the Apennines

But Apennine ecology is only part of the explanation. To make sense of the long series of floods that afflicted the Liri Valley in the nineteenth century we need to consider them as socio-natural disasters. While neither the Bourbon, the French nor the Savoy State prioritised efforts to drain the marshes and reforest the slopes of southern Italy, what did receive priority meanwhile was the politics of enclosure and privatisation of the country's land and water. And the environmental costs of those politics were long borne by the poor and voiceless inhabiting the inland slopes of the Apennines.

Epilogue

Common Waters

I

This book has narrated the environmental transformation of a Mediterranean valley, through the eyes of the educated middle classes of southern Italy – those who left written documents concerning their vision, use and misuse of water in the age of industry. The story has focussed on the cultural and political changes that led to the remaking of a river as enclosed property; and on the consequences that such transformation had on people's working and living conditions. I have argued that early Political Economy ideas about the evils of common property and the coincidence between private interest and the public good greatly contributed to environmental degradation and increased flood risk; and that the same set of assumptions also caused State policies aimed at disciplining the disorderly use of water to fail.

In the Liri Valley story, starting with the abolition of feudality, land and water were appropriated by the rural bourgeoisie, who used them in purely individualistic fashion, with no consideration of social and environmental costs. Both agriculturalists and industrialists contributed to the palpable increase of environmental risk during the nineteenth century, causing river siltation, inundations, unhealthy work and living conditions and malaria. In addition, they altered the river ecosystem in a way incompatible with fish and aquatic life: though the available sources contain only scant and unsystematic observations on such matters, in 1852 the writer Filippo Cirelli could already notice how three fish species – Brown Trout *(Salmo trutta)*, Goatfish *(Barbus plebejus)* and Rovella *(Rutilus rubilio)* – were disappearing from the Liri due to both excessive canalisation and industrial pollution. Conversely, he registered the abundance of Planer's Lamprey *(Petromyzon planeri)* and River Lamprey *(Lampetra fluviatilis)*, neither of which can be found in the Liri River today. A number of fish, among them barbells, squami anthias, round sardinellas, grey mullets, sturgeons, velias, eels and shrimps (all mentioned in the *Statistica Murattiana* of 1811) are also missing from those found in the Liri Valley by a recent biodiversity survey. All these species have probably diminished in a manner directly proportional to the addition of water plugs and diversions and noticeably so with the advent of the hydroelectric engine, local biologist Rosaria Bellucci notes. Conversely, other species of fish inhabit the Liri River today: they have been introduced for sport fishing in more or less recent decades, mostly selected from non-spawning species tolerant of pollution. Some of these have been predating local species, so further incrementing their disappearance[379]. Moreover, the river otter, also mentioned in the *Statistica*, has now vanished from the Liri Valley, as it has from most other Italian rivers.

Paradoxically, this new economy of water was created out of a powerful mix of political economy and ecological ideas, according to which river degradation in southern Italy was the most important aspect of economic backwardness, the two linked in a vortex of circular causation. Breaking this vicious circle and restoring the reputed natural wealth of the country's land required the destruction of both feudal rule and of customary access to nature and the advent of private property. This was the myth of origins for the spread of capitalism in southern Italy: coming after centuries of feudal oppression and waste of resources, the new relations of production would rescue both the country's environment and economy from decline. Only the political economy of private property – so the theory went – could sort out the 'disorder of water'.

Much of this book's narrative has thus dealt with the problem of explaining the key concept of water 'disorder', a concept not unique to the Neapolitan philosopher and civil servant of the time, but certainly typical of a representation of nature as a stable, balanced and orderly system – a vision that has been thoroughly shaken in the last few decades[380]. Today's ecologists tend to read watersheds in terms of a dialectic interaction between ecological 'integrity' and 'disturbance'. River ecosystems should be left alone as much as possible, modern hydrology tells us. Riverbeds should be neither channelled nor embanked and water should be left free to periodically overflow the natural banks and inundate the floodplain. This should be so for a number of reasons concerning the health of the ecosystem and the effective protection of humans from environmental risk. In short, if we don't want to be inundated, all we need to do is to stay out of the floodplain[381].

Modern hydrology seems more ecologically sound than eighteenth century natural philosophy or nineteenth century hydraulic engineering. And yet, judging the Liri River story with the eyes of a modern river ecologist would be highly misleading. First, and obviously, because many things about ecological interconnections were unknown and were first realised only after the industrial transformation of the valley was an accomplished fact[382]. Second, and more importantly, because the origins of river degradation and habitat loss have to be researched on a much wider scale, that of European enclosure-and-improvement culture, as well as that of nation-state building and the struggle of European dynasties for political and economic supremacy over the world-economy of the time. Local actors fit into this larger scenario, without which their choices could not be properly understood. The Improvement project acted as a powerful force of socio-environmental change, just like so-called globalisation and European Union directives act today in orienting environmental and economic policies on the local scale. In fact, today's river restoration projects could not be properly understood without considering the de-industrialisation and de-materialisation of the economy in first world countries, of which the Liri Valley is now part.

II

It is not yet clear what will come next in the Liri Valley. With most of the factory jobs gradually having disappeared in the course of the past half century and thousands of families having moved out of the area, no new socio-natural order is in sight to replace the disorder of industry. Great plans have been made for Isola Liri to reinvent itself as a site for tourism and services based on a mixture of natural and historical attractions: remnants of industrial archaeology included in a 'fluvial and technology park', centred on the waterfall–palace complex. This 'urban regeneration' project, a joint public–private venture co-financed by the European Union, will allow the restoration of many old factories lining the riverbank and their reuse as hotels, sites for computer and telecommunication companies, and for cultural and recreation activities (including a ticket office located in the old slaughterhouse)[383].

Like all restoration projects, this one too is crucially based on history[384]. What comes to be reused, in fact, is not just old buildings and urban space, but the past itself. Or at least one particular vision of the past, based on uncritical and celebratory views of the industrial era, as the time in which 'urban life was organised around industry', and people 'adapted their bio-rhythm to that of the textile and paper productions that gave prosperity to the town'[385].

Being a project under construction and with many things still not entirely solid, it is still possible to formulate some questions about the future. Will this project (like many restoration–conservation schemes to date), with its massive input of capital and technocracy, produce new forms of exclusion, leaving out alternative visions for a different reconnection between local people and the river? Will other views of the past, centred on environmental costs and on those who paid them, be taken into account in the official collective memory of Isola's fluvial park, or will this just be a celebration of social power, technology, entrepreneurship and the literary landscape? Will this collective memory include the trout, the otter and the other disappearing species of the Liri watershed and will they have a chance to find their way back to the river? Will the 'fluvial park' be an opportunity for a new environmentalism to emerge in the Liri Valley, one more aware of the unequal power relationships and forms of exclusion that were inscribed in past landscapes? Finally, will the 'fluvial park' be a chance to reverse the fundamental alienation of local people from their river, or will this continue to be the toll they pay to present-day reconfigurations of the 'economy of water'?

Similar questions can be raised concerning the inclusion–exclusion dialectic on the broader scale, that of the Watershed Authority. At this level, a technocratic environmental planning takes shape and undertakes continuous formal exchanges with territorial planning bureaus at the regional, district and town level. But just how are citizens expected to participate in this techno-bureaucratic scheme and to express their vision (or non-vision) and experience of the river; how might locally based, or just different views of water–society relationships fit into this process of watershed planning; and, more importantly, out of what ideas of nature and what social relationships are the new improving schemes of the post-industrial era imagined, negotiated and eventually imposed on the local space?

III

Perhaps needless to say, it has not been the intention of this book to advocate a return to the nature–society relationships typical of the pre-capitalist rural economy – a world already marked by unequal and oppressive relationships of resource access and exploitation. The very idea of restoring a past harmony with nature has been clearly exposed and critiqued throughout the book, as historical experience has shown how such ideas often bring about oppressive and unequal social relationships. On the contrary, the book has aimed to contribute to the intellectual and political efforts of countless individuals and groups to think both ecology and society in new ways. In fact, to 'save' the Liri Valley – that is to have at last a sustainable and socially equitable interaction with the river – we might need a new kind of political economy: one that helps us to see water as neither private or public, but as belonging to the sphere of the common; and – perhaps even more challenging – to develop a more comprehensive idea of the 'common', aware of social inequalities and inclusive of the non-human living world.

Bibliography

Agrawal, A. and K. Sivaramakrishnan (eds.) 2001. *Agrarian Environments: Resources, Representations, and Rule in India*. Durham: Duke University Press.

Andretta, Marzia. 2008. *Il meridionalista. Giustino Fortunato e la rappresentazione del Mezzogiorno*. Roma: XL Edizioni.

Archivio di Stato di Frosinone. 1992. *Viabilità e territorio nel Lazio meridionale. Persistenze e mutamenti fra '700 e '800*. Frosinone: Don Guanella.

Armiero, M. 1999. *Il territorio come risorsa. Comunità economie istituzioni nei boschi d'Abruzzo*. Napoli: Liguori

Armiero, M. 2006. 'Italian Mountains', in M. Armiero (ed.) *Views from the South. Environmental Stories from the Mediterranean World*. Napoli: CNR.

Armiero, M. 2008. 'Seeing Like a Protester. Nature, Power, and Environmental Struggles'. *Left History*, 1, 59–76.

Armiero, M. and M. Hall. 2004. 'Italy', in S. Krech, J.R. McNeill and C. Merchant (eds.) *Encyclopedia of World Environmental History*. New York: Routledge.

Armiero, M. and W. Palmieri. 2002. 'Boschi e rivoluzioni nel Mezzogiorno. La gestione, gli usi e le strategie di tutela nelle congiunture di crisi di regime (1799-1860)', in A. Lazzarini (ed.), *Processi di diboscamento montano e politiche territoriali. Alpi e Appennini dal settecento al duemila*. Milano: F. Angeli.

Armiero, M. and N. Peluso. 2008. 'Insurgent Natures and Nations: Unpacking Socio-Environmental Histories of Forest Conflict', paper presented at the conference *Common Ground, Converging Gazes. Integrating the Social and Environmental in History*, École des Hautes Études en Sciences Sociales, Paris, France, 13 September 2008; http://cnr-it.academia.edu/MarcoArmiero/Papers.

Arnold, David. 2006. *The Tropics and the Traveling Gaze: India, Landscape, and Science, 1800-1856*. Seattle: University of Washington Press.

Astarita, Tommaso. 2005. *Between Salt Water and Holy Water. A History of Southern Italy*. New York: W.W. Norton & Co.

Barca, S. 2004. 'Il capitale naturale. Acque e rivoluzione industriale in valle del Liri'. *Memoria e ricerca*, 15.

Barca, S. 2006. 'Running Italian Waters. Hydrology and Water Law in the Age of Industrialization', in M. Armiero (ed.) *Views from the South. Environmental Stories from the Mediterranean World*. Napoli: Consiglio Nazionale delle Ricerche.

Bellucci, R. 2009. *Biodiversità nel parco fluviale di Isola del Liri*. Frosinone: Edizioni LEA.

Benoit, S. 1985. 'De l'hydromécanique à l'hydro-électricité', in F. Cardot (ed.) *La France des electriciens, 1880-1980*. Paris: Presses Universitaires de France.

Berg, Maxine. 1980. *The Machinery Question and the Rise of Political Economy, 1815-1848*. Cambridge and New York: Cambridge University Press.

Bevilacqua, P. 1989. 'Le rivoluzioni dell'acqua. Irrigazioni e trasformazione dell'agricoltura tra sette e novecento', in P. Bevilacqua (ed.) *Storia dell'agricoltura Italiana in età contemporanea*, vol. I. *Spazi e paesaggi*. Venezia: Marsilio.

Bevilacqua, Piero. 1996. *Tra natura e storia. Ambiente, economie, risorse in Italia*. Roma: Donzelli.

Bevilacqua, Piero. 1997. *Breve storia dell'Italia meridionale: dall'ottocento a oggi*. Roma: Donzelli.

Bevilacqua, P. 2000. 'Contexts and Debates: Environmental Intervention and Water Resource Management in the History of the Mezzogiorno'. *Modern Italy*, 1, 63–71.

Bevilacqua, P. 2005. 'I caratteri originali della storia ambientale Italiana. Proposte di discussione'. *I Frutti di Demetra*, 8.

Bevilacqua, P. 2009 [1998]. *Venice and the Water. A Model for Our Planet*. Solon, ME: Polar Bear and Co.

Bevilacqua, P. and M. Rossi Doria (eds.) 1984. *Le bonifiche in Italia dal '700 ad oggi*. Bari: Laterza.

Bianchini, Ludovico. 1971 [1857]. *Storia delle finanze del regno delle due Sicilie*. Napoli: Edizioni Scientifiche Italiane.

Bigatti, Giorgio. 1995. *La provincia delle acque: ambiente, istituzioni e tecnici in Lombardia tra sette e ottocento*. Milan: F. Angeli.

Bigatti, G. (ed.) 2000. *Uomini e acque. Il territorio Lodigiano tra passato e presente*. Lodi: Giona.

Black, Jeremy. 2003. *Italy and the Grand Tour*. New Haven: Yale University Press.

Blackbourn, David. 2006. *The Conquest of Nature. Water, Landscape and the Making of Modern Germany*. New York and London: W.W. Norton.

Bloch, Marc. 1967. *Land and Work in Medieval Europe*. Berkeley: University of California Press

Brancaccio, G. 1988. 'La raffigurazione della Campania e del Molise nella cartografia Napoletana del secolo XVIII', in E. Narciso (ed.) *Illuminismo meridionale e comunità locali*. Napoli: Guida.

British Museum (Natural History) 1904. *The History of the Collections Contained in the Natural History Departments of the British Museum*. London: Longman & Co.

Brüggemeier, F.J. 1994. 'A Nature Fit for Industry: The Environmental History of the Ruhr Basin'. *Environmental History Review*, 18, 35–54.

Carbone, Arduino. 1971. *Giustiniano Nicolucci e la sua patria*. Isola del Liri: Comune di Isola Liri.

Casey, Edward. 1997. *The Fate of Place: A Philosophical History*. Berkeley: University of California Press.

Castonguay, S. 2007. 'The Production of Flood as Natural Catastrophe: Extreme Events and the Construction of Vulnerability in the Drainage Basin of the St. Francis River (Quebec), Mid-nineteenth to Mid-twentieth Century'. *Environmental History*, 12, 820–44.

Cazzola, F. 2006. 'Sui caratteri originali della storia ambientale italiana'. *I Frutti di Demetra*, 11.

Cencelli, Alberto. 1920. *La proprietà collettiva in Italia*. Milan: Hoepli.

Cimmino, C. 1986a. 'Capitalismo e classe operaia nel Mezzogiorno', in C. Cimmino (ed.) *Economia e società in valle del Liri nel secolo XIX. L'industria laniera*. Caserta: Istituto per la Storia del Risorgimento Italiano.

Bibliography

Cimmino, C. 1986b. 'La vendita dei beni dell'asse ecclesiastico', in C. Cimmino (ed.) *Economia e società nella valle del Liri nel sec. XIX*. Caserta: Istituto per la Storia del Risorgimento Italiano.

Cioc, Mark. 2002. *The Rhine: An Eco-Biography, 1815–2000*. Seattle: University of Washington Press.

Cioc, M. 2004. 'The Political Ecology of the Rhine'. in C. Mauch (ed.) *Nature in German History*. New York and Oxford: Berghahn Books.

Ciriacono, S. (ed.) 1998. *Land Drainage and Irrigation*. Aldershot and Brookfield, VT: Ashgate/Variorum.

Ciriacono, Salvatore. 2006a. *Building on Water: Venice, Holland, and the Construction of the European Landscape in Early Modern Times*. New York: Berghahn Books.

Ciriacono, S. 2006b. 'Hydraulic Energy, Society and Economic Growth', in Dooley, B. (ed.) *Energy and Culture. Perspectives on the Power to Work*. Burlington, VT: Ashgate.

Ciriacy-Wantrup, S.V. and R.C. Bishop. 1975. '"Common Property" as a Concept in Natural Resource Policy'. *Natural Resources Journal*, 15, 177–85.

Corona, Gabriella. 1994. *Demani e individualismo agrario nel regno di Napoli 1780–1806*. Napoli: ESI.

Corona, G. 2004. 'Stato, Proprietà Privata e Possesso Collettivo', in I. Zilli (ed.) *Lo stato e l'economia tra restaurazione e rivoluzioni*, vol. I. Napoli: ESI.

Corradini, Ferdinando. 2004. *Di Arce in Terra di Lavoro*, vol. II. Arce: Amminstrazione Comunale.

Cosgrove, Denis. 1998. *Social Formation and Symbolic Landscape*. Madison: University of Wisconsin Press.

CRGC (Corpo Reale del Genio Civile). 1907. *Misurazione delle portate dei corsi d'acqua*. Roma: Tipografia e Zecca dello Stato.

Cronon, W. 1992. 'A Place for Stories. Nature, History and Narrative'. *Journal of American History*, 3, 1347–76.

Cronon, W. 1996. 'The Trouble with Wilderness; or, Getting Back to the Wrong Nature', in W. Cronon (ed.) *Uncommon Ground. Rethinking the Human Place in Nature*. New York: W.W. Norton.

Crosby, Alfred. 2006. *Children of the Sun. A History of Humanity's Unappeasable Appetite for Energy*. New York and London: W.W. Norton & Company.

Cuciniello, Antonio and Lorenzo Bianchi (eds.) 1971 [1829]. *Viaggio pittorico nel regno delle due Sicilie dedicato a sua maestà il Re Francesco Primo*. Napoli: SEM.

D'Elia, C. (ed.) 1992. *Il Mezzogiorno agli inizi dell'ottocento*. Roma-Bari: Laterza.

D'Elia, Costanza. 1994. *Bonifiche e stato nel Mezzogiorno (1815-1860)*. Napoli: ESI.

D'Elia, Costanza. 1996. *Stato padre, stato demiurgo. I lavori Pubblici nel Mezzogiorno*. Bari: Edipuglia.

D'Souza, Rohan. 2006 *Drowned and Dammed. Colonial Capitalism and Flood Control in Eastern India*. New Delhi: Oxford University Press.

Davis, John Anthony. 1979. *Società e imprenditori nel regno Borbonico*. Bari: Laterza.

Bibliography

Davis, J.A. 1994. 'The Mezzogiorno and Modernisation: Changing Contours of Public and Private during the French Decennio', in P. Macry and A. Massafra (eds.) *Fra storia e storiografia. Scritti in onore di Pasquale Villani*. Bologna: Il Mulino.

Davis, John Anthony. 2006. *Naples and Napoleon. Southern Italy and the European Revolutions (1780-1860)*. Oxford and New York: Oxford University Press.

De Lorenzo, Renata. 2001a. 'Accademismo e Associazionismo tra "Desideri" Riformistici E "Passioni" Giacobine: Carlo Lauberg', in R. De Lorenzo. *Un regno in bilico. Uomini, eventi e Luoghi nel Mezzogiorno preunitario*. Roma: Carocci.

De Lorenzo, R. 2001b. 'Tradizione e innovazione: "Uomini di scienza" e rivoluzione in Terra di Bari e Basilicata', in R. De Lorenzo. *Un regno in bilico. Uomini, eventi e luoghi nel Mezzogiorno Preunitario*. Roma: Carocci.

De Majo, Silvio. 1990. *L'industria protetta. Lanifici e cotonifici in Campania nell'ottocento*. Napoli: Athena.

De Marco, D. (ed.) 1988. *La "Statistica" del regno di Napoli nel 1811*, Roma: Accademia Nazionale dei Lincei.

De Matteo, Luigi. 1984. *Governo, credito e industria laniera nel Mezzogiorno: da Murat alla crisi post-unitaria*. Napoli: nella sede dell'Istituto Italiano per gli Studi Filosofici.

De Negri, F. 1992. 'La "reintegra" al demanio dello stato di Sora. Un momento del dibattito sulla feudalità nel regno di Napoli alla fine del '700', in Archivio di Stato di Frosinone. *Viabilità e Territorio nel Lazio Meridionale*. Frosinone: Don Guanella.

De Rivera, Carlo Afan. 1825. *Memoria intorno alle devastazioni prodotte dalle acque a cagion de' diboscamenti*. Napoli: Reale Tipografia della Guerra.

De Rivera, Carlo Afan. 1832–42. *Considerazioni sui mezzi da restituire il valore proprio a' doni che ha la natura largamente conceduto al regno delle Due Sicilie*. Napoli: dalla Stamperia e Cartiera del Fibreno.

De Rosa, L. 1979. 'Property Rights, Institutional Change and Economic Growth in Southern Italy in the XVIIIth and XIXth Centuries'. *Journal of European Economic History*, 3, 531–2.

De Sanctis, Riccardo. 1986. *La nuova scienza a Napoli tra '700 e '800*. Roma: Laterza.

De Seta, C. 1982. 'L'Italia nello specchio del Grand Tour', in De Seta (ed.) *Storia d'Italia. Annali 5. Il paesaggio*. Torino: Einaudi.

Debeir, Jean-Claude, Jean-Paul Déleage and Claude Émery. 1991 [1986]. *In the Servitude of Power: Energy and Civilization through the Ages*. London: Zed Books.

Dell'Orefice, A. 1988. 'L'industria della carta nel XIX secolo', in *Trasformazioni industriali nella media valle del Liri in età moderna e contemporanea*. Isola de Liri: Rotary Club.

Desrosières, Alain. 1998. *The Politics of Large Numbers. A History of Statistical Reasoning*. Cambridge MA: Harvard University Press.

Dewerpe, Alain. 1986. 'Crescita e ristagno protoindustriali nell'Italia meridionale: la valle del Liri', in A. De Clementi (ed.) *La società inafferrabile*. Roma: Editori Riuniti.

Di Biasio, Aldo. 1997. *Territorio e viabilità nel Lazio meridionali: gli antichi distretti di Sora e di Gaeta. 1800–1860*. Marina di Minturno: Caramanica.

Di Biasio, Aldo. 2004. *Politica e amministrazione del territorio nel Mezzogiorno d'Italia tra settecento e ottocento*. Napoli: ESI.

Bibliography

Dias, Francesco. 1841. *Legislazione positiva del regno delle Due Sicilie dal 1806 a tutto il 1840.* Napoli: dalla Stamperia e Cartiera del Fibreno.

Dickie, John. 1999. *Darkest Italy: The Nation and Stereotypes of the Mezzogiorno, 1860-1900.* London: Palgrave Macmillan.

Fazzini, Antonio. 1836. 'Isola di Sora'. *Poliorama pittoresco,* vol. I.

Federico, G. and P. Malanima. 2004. 'Progress, Decline, Growth: Product and Productivity in Italian Agriculture, 1000–2000'. *Economic History Review,* 3, 437–64.

Ferri, M. 1986. 'Valle del Liri: la grande porta del brigantaggio meridionale', in C. Cimmino (ed.) *Economia e società nella valle del Liri nel sec. XIX.* Caserta: Istituto per la Storia del Risorgimento Italiano.

Finotti, Giuseppe. 1891. *Monografia industriale e commerciale di Terra di Lavoro.* Caserta: Camera di Commercio e Industria.

Fischer-Kowalski, Marina and Helmut Haberl (eds.) 2007. *Socioecological Transitions and Global Change: Trajectories of Social Metabolism and Land Use.* Cheltenham: Edward Elgar.

Flores, D. 1996. 'Place: An Argument for Bioregional History'. *Environmental History,* 4, 1–18.

Foster, John Bellamy. 1999. *The Vulnerable Planet: A Short Economic History of the Environment.* New York: Monthly Review Press.

Founders Society. 1975. *French Painting 1774-1830, the Age of Revolution.* Detroit: Wayne State University Press.

Furet, F. 1989. 'Feudal System' in F. Furet and M. Ozouf (eds.) *A Critical Dictionary of the French Revolution.* Cambridge: Harvard University Press.

Galanti, Giuseppe Maria. 1969 [1794]. *Della descrizione geografica e politica delle Sicilie.* (F. Assante and D. Demarco eds.) Napoli: Edizioni Scientifiche Italiane.

Gallo, L. 2002. 'Ambiente e paesaggio in Magna Grecia. Le fonti letterarie', in *Ambiente e paesaggio nella Magna Grecia.* Taranto: Istituto per la storia e l'archeologia della Magna Grecia.

Gambi, L. 1972. 'I valori storici dei quadri ambientali', in *Storia d'Italia,* vol. 1. *I caratteri originali.* Torino: Einaudi.

Gaspari, O. 2000. 'Questione montanara e questione meridionale', in P. Bevilacqua and G. Corona (eds.) *Ambiente e risorse nel Mezzogiorno contemporaneo.* Corigliano Calabro: Meridiana Libri.

Genovesi, Antonio. 1977. *Scritti* (F. Venturi ed.) Torino: Einaudi.

Gordon, R. 1983. 'Cost and Use of Water Power During Industrialisation in New England and Great Britain: a Geological Reinterpretation'. *Economic History Review,* 2, 240–59.

Gordon, R. 1985. 'Hydrological Science and the Development of Waterpower for Manufacturing'. *Technology and Culture,* 2, 204–35.

Greenberg, D. 1982. 'Reassessing Power Patterns of the Industrial Revolution: An Anglo-American Comparison'. *The American Historical Review,* 5, 1237–61.

Greenberg, D. 1990. 'Energy, Power and Perceptions of Social Change in the Early Nineteenth Century'. *The American Historical Review,* 3, 693–714.

Grossi, Paolo. 1981 [1977]. *An Alternative to Private Property: Collective Property in the Juridical Consciousness of the Nineteenth Century* (trans. Lydia G. Cochrane). Chicago: University of Chicago Press.

Guha, Ramachandra. 2000. *Environmentalism: A Global History*. Oxford and New Delhi: Oxford University Press.

Gutwirth, S. (ed.) 1978. *Jean Joseph-Xavier Bidauld (1758-1846). Peintures et dessins*. Nantes: Chiffoleau.

Hall, Marcus. 2005. *Earth Repair: A Transatlantic History of Environmental Restoration*. Charlottesville: University of Virginia Press.

Hannerz, Ulf. 1996. *Transnational Connections: Culture, People, Places*. London: Routledge.

Hardin, G. 1968. 'The Tragedy of the Commons'. *Science*, 162, 1243–8.

Harris, Diana. 2003. *The Nature of Authority. Villa Culture, Landscape and Representation in Eighteenth-Century Lombardy*. University Park, PA: The Pennsylvania State University Press.

Harvey, David. 1989. *The Condition of Postmodernity. An Enquiry into the Origins of Cultural Change*. Oxford: Basil Blackwell.

Harvey, David. 1996. *Justice, Nature and the Geography of Difference*. Cambridge, MA: Blackwell Publishers.

Harvey, M. 2008. 'Interview with Donald Worster'. *Environmental History*, 13, 140–55.

Henneberg, M. and R.J. 1998. 'Biological Characteristics of the Population Based on Analysis of Skeletal Remains', in J. Carter (ed.) *The Chora of Metaponto. The Necropoloeis*. Austin: University of Texas Press.

Hills, Richard. 1970. *Power in the Industrial Revolution*. Manchester: Manchester University Press.

Hobsbawm, Eric. 1990. *Echoes of the Marseillaise. Two Centuries Look Back on the French Revolution*. New Brunswick: Rutgers University Press.

Hobsbawm, Eric. 1996 [1962]. *The Age of Revolution 1789-1848*. New York: Vintage Books.

Hobsbawm, Eric. 2000 [1969]. *Bandits*. New York: Norton & Co.

Holmes, Douglas. 1989. *Cultural Disenchantments. Worker Peasantries in Northeast Italy*. Princeton, NJ: Princeton University Press.

Hunter, Louis C. 1979. *A History of Industrial Power in the United States*, vol. 1. *Waterpower*. Charlottesville: University Press of Virginia.

Isenburg, T. 2000. 'Conoscenza e governo dei fiumi. A proposito del Po', in G. Bigatti (ed.). *Uomini e acque. Il territorio Lodigiano tra passato e presente*. Lodi: Giona.

Isernia, O. 1986. 'La vendita dei beni demaniali nella valle del Liri dopo l'unità' in C. Cimmino (ed.) *Economia e società nella Valle del Liri nel sec. XIX*. Caserta: Istituto per la Storia del Risorgimento Italiano.

Jones, Eric L. 1974. *Agriculture and the Industrial Revolution*. Oxford: Blackwell.

Kapp, William. 1971[1950]. *The Social Costs of Private Enterprise*. New York: Schocken Books.

Kelman, Ari. 2003. *A River and its City: the Nature of Landscape in New Orleans*. Berkeley: University of California Press.

Krausmann, F. H. Schandl, R.P. Sieferle. 2008. 'Socio-ecological Regime Transitions in Austria and the United Kingdom'. *Ecological Economics,* 65, 187–201.

Kriedte, P. H. Medick and J. Schlumbohm (eds.) 1981 [1977]. *Industrialisation before Industrialisation.* Cambridge: Cambridge University Press.

Landes, David. 2003 [1969]. *The Unbound Prometheus. Technical Change and Industrial Development in Western Europe from 1750 to the Present.* Cambridge and New York: Cambridge University Press.

Lauri, Achille. 1914. *Sora, Isola Liri e Dintorni.* Sora: Vincenzo D'Amico.

Lumley, R. and J. Morris (eds.) 1997. *The New History of the Italian South: The Mezzogiorno Revisited.* Exeter, UK: University of Exeter Press.

Macciocchi, Maria Antonietta. 1993. *Cara Eleonora.* Milano: Rizzoli.

Macry, Paolo. 1974. *Mercato e società nel regno di Napoli.* Napoli: Guida.

Macry, Paolo. 2002 [1988]. *Ottocento: famiglia, elites e patrimoni a Napoli.* Bologna: Il Mulino.

Macry, P. (ed.) 2003. *Quando crolla lo stato: Studi sull'Italia preunitaria.* Napoli: Liguori.

MAIC (Ministero di Agricoltura, Industria e Commercio). 1895. *Carta idrografica d'Italia. Liri-Garigliano, Paludi Pontine e Fucino.* Roma: Tipografia Nazionale.

Malanima, P. 2003a. 'Energy Systems in Agrarian Societies: The European Deviation', in *Economia e Energia Secc. XIII-XVIII.* Florence: Prato International Institute of Economic History.

Malanima, P. 2003b. 'Measuring the Italian Economy: 1300–1861' *Rivista di Storia Economica.*

Mancini, M. 1882. 'Sulle condizioni agrarie del circondario di Sora', in *Atti della giunta per la inchiesta agraria e sulle condizioni della classe agricola.* Roma: Tip. Forzani & C.

Mantoux, Paul. 1999 [1909]. *La rivoluzione industriale.* Roma: Editori Riuniti.

Marx, Leo. 1964. *The Machine in the Garden. Technology and the Pastoral Ideal in America.* New York: Oxford University Press.

Mathias, P. 1999. 'Dalla Ruota Idraulica alla Macchina a Vapore', in V. Castronovo (ed.) *Storia dell'economia mondiale* vol. 3. *L'età della rivoluzione industriale.* Roma and Bari: Laterza.

Mauch, C. 2004. 'Introduction. Nature and Nation in Transatlantic Perspective', in C. Mauch (ed.) *Nature in Germany History.* New York and Oxford: Berghahn Books.

McCay, B. and J. Acheson. 1987. *The Question of the Commons: The Culture and Ecology of Communal Resources.* Tucson: University of Arizona Press.

McEvoy, Arthur. 1986. *The Fisherman's Problem: Ecology and Law in the California Fisheries. 1850-1980.* Cambridge and New York: Cambridge University Press.

McNeill, John R. 1992. *The Mountains of the Mediterranean World: An Environmental History.* Cambridge and New York: Cambridge University Press.

McNeill, J.R. 2000. *Something New Under the Sun. An Environmental History of the Twentieth-Century World.* New York: Norton.

Meiksins Wood, Ellen. 1999. *The Origin of Capitalism.* New York: Monthly Review Press.

Merchant, Carolyn. 1989. *Ecological Revolutions. Nature, Gender, and Science in New England.* Chapel Hill: University of North Carolina Press.

Merchant, C. 1994. 'Introduction' to C. Merchant (ed.) *Key Concepts in Critical Theory: Ecology*. New York: Humanity Books.

Merchant, Carolyn. 2004. *Reinventing Eden. The Fate of Nature in Western Culture*. New York and London: Routledge.

Migliore, D.N. 2006. 'Il catasto provvisorio', in I. Ascione and A. Di Biasio (eds.) *Caserta al tempo di Napoleone. Il decennio Francese in Terra di Lavoro*. Napoli: Electa.

Mitchell, T. 2000. 'The Stage of Modernity', in T. Mitchell. (ed.) *Questions of Modernity*. Minneapolis: University of Minnesota Press.

Moe, Nelson. 2002. *The View from Vesuvius. Italian Culture and the Southern Question*. Berkeley: University of California Press.

Monticelli, Teodoro. 1809. *Memoria sull'economia delle acque da ristabilirsi nel regno di Napoli*. Napoli: Stamperia Reale.

North, Douglas C. and Robert Thomas. 1973. *The Rise of the Western World. A New Economic History*. Cambridge: Cambridge University Press.

Nye, David. 2005. *America as Second Creation. Technology and Narratives of New Beginnings*. Cambridge, MA and London: The MIT Press.

O'Connor, J. 1998. *Natural Causes. Essays in Ecological Marxism*. New York and London: Guilford Press.

Orsi, Jared. 2004. *Hazardous Metropolis: Flooding and Urban Ecology in Los Angeles*. Berkeley: University of California Press.

Ostrom, Elinor. 1993. 'Coping with Asymmetries in the Commons: Self-Governing Irrigation Systems Can Work'. *Journal of Economic Perspectives*, 4, 93–112.

Ottani Cavina, Anna. 2004. *Geometries of Silence. Three Approaches to Neoclassical Art*. Columbia U.P: New York.

Paavola, J. 2002. 'Water Quality as Property: Industrial Water Pollution and Common Law in the Nineteenth Century United States'. *Environment and History*, 8, 295–318.

Palmieri, W. 1993. 'Introduzione' in W. Palmieri (ed.) *Il Mezzogiorno agli inizi della restaurazione*. Bari: Laterza.

Palmieri, W. 2000. 'Il bosco nel Mezzogiorno preunitario tra legislazione e dibattito', in P. Bevilacqua and G. Corona (eds.) *Ambiente e risorse nel Mezzogiorno contemporaneo*. Corigliano Calabro: Meridiana Libri.

Palmieri, W. 2010. 'Nature, Culture, and Disaster: Vesuvius and the Nola Mudslides of the Nineteenth Century', in M. Armiero and M. Hall (eds.) *Nature and History in Modern Italy*. Athens: Ohio University Press.

Palumbo, Manfredi. 1979. *I comuni meridionali prima e dopo le leggi eversive della feudalità*, Bologna: Forni.

Parisi, Roberto and Adriana Pica. 1996. *L'impresa del Fucino. Architettura delle acque e trasformazione ambientale nell'età dell'industrializzazione*. Napoli: Athena.

Patriarca, Silvana. 1996. *Numbers and Nationhood. Writing Statistics in 19th Century Italy*. Cambridge and New York: Cambridge University Press.

Bibliography

Peets, R. and M. Watts. 1996. 'Towards a Theory of Liberation Ecology', in R. Peets and M. Watts (eds.) *Liberation Ecologies. Environment, Development, Social Movements.* London and New York: Routledge.

Peluso, N. and M. Watts (eds.) 1998. *Violent Environments.* Ithaca and London: Cornell University Press.

Petrusewicz, Marta. 1996 [1989]. *Latifundium: Moral Economy and Material Life in a European Periphery.* Ann Arbor: The University of Michigan Press.

Petrusewicz, Marta. 1998. *Come il meridione divenne una questione. Rappresentazioni del sud prima e dopo il quarantotto.* Soveria Mannelli: Rubbettino.

Pistilli, Ferdinando. 1824. *Descrizione storico-filologica delle antiche, e moderne città e castelli esistenti accosto de' fiumi Liri e Fibreno.* Napoli: Stamperia Francese.

Platt, H. 2002. The 'Hardest Worked River'. The Manchester Floods and the Industrialization of Nature', in G. Massard-Guilbaud and D. Schott (eds.) *Cities and Catastrophes. Coping with Emergency in European History.* Frankfurt am Main: Peter Lang.

Platt, Harold L. 2005. *Shock Cities: The Environmental Transformation and Reform of Manchester and Chicago.* Chicago: University of Chicago Press.

Polanyi, Karl. 1944. *The Great Transformation. The Political and Economic Origins of Our Time.* Boston: Beacon Press.

Pollard, Sidney. 1997. *Marginal Europe. The Contribution of Marginal Lands since the Middle Ages.* New York: Oxford University Press.

Prigogine, Y. 1994. 'Science in a World of Limited Predictability', in Merchant, C. (ed.) *Key Concepts in Critical Theory: Ecology.* New York: Humanity Books.

Protasi, Maria R. 2002. *Operai e contadini della valle del Liri. Condizioni di vita, famiglia, lavoro (1860–1915).* Sora: Centro Studi Sorani "Vincenzo Patriarca".

Rabinow, P. and N. Rose (eds.) 2003. *The Essential Foucault: Selections from Essential Works of Foucault, 1954-1984.* New York: New Press.

Rackham, O. 2006. 'Mountains, woods, and waters in the European Mediterranean: a summary for the last 200 years', in M. Armiero (ed.) *Views from the South. Environmental Stories from the Mediterranean World.* Napoli: CNR.

Radkau, Joakim. 2008 [2002]. *Nature and Power: A Global History of the Environment.* Cambridge and New York: Cambridge University Press.

Raimondo, Sergio. 2000. *La risorsa che non c'è più. Il lago del Fucino dal 16° al 19° secolo.* Manduria: Lacaita.

Ramazza, S. 1996. 'L'organizzazione territoriale dell'amministrazione delle acque nello Stato italiano'. *Storia urbana*, 74.

Ramella, Francesco. 1983. *Terra e telai: sistemi di parentela e manifattura nel biellese dell 'ottocento.* Torino: Einaudi.

Reynolds, Terry S. 1983. *Stronger than a Hundred Men. A History of the Vertical Water Wheel.* Baltimore: Johns Hopkins University Press.

Rizzello, Marcello and Daniela Campagna. 1988. *L'archivio storico comunale di Isola del Liri.* Sora: Pasquarelli.

Robertson, J. 1997. 'The Enlightenment above National Context: Political Economy in Eighteenth-Century Scotland and Naples'. *The Historical Journal*, 3, 667–697.

Robertson, J. 2000. 'Enlightenment and Revolution: Naples 1799'. *Transactions of the Royal Historical Society*, 10, 17–44.

Romanelli, Domenico. 1819. *Viaggio da Napoli a Montecassino ed alla celebre cascata d'acqua nell'Isola di Sora*. Napoli: A. Trani.

Romano, Antonia. 2005. *Storia di una rete. Famiglia, professione e politica nel carteggio di Antonio Ranieri (1855-1865)*. Unpublished dissertation, University of Napoli "Federico II", Department of History, PhD Programme in European History, 2004-05 (http://www.fedoa.unina.it/300/01/tesidottromano.pdf).

Royal Society of London. 1870. *Catalogue of Scientific Papers, vol. IV (1800-1863)*. London: Eyre and Spottiswoode.

Saikku, Mirko. 2005. *This Delta, This Land: An Environmental History of the Yazoo–Mississippi Floodplain*. Athens: University of Georgia Press.

Salvemini, Biagio. 1981. *Economia politica e arretratezza meridionale nell'età del risorgimento. Luca de Samuele Cagnazzi e la diffusione dello Smithianesimo nel regno di Napoli*. Lecce: Milella.

Salvemini, B. 2000. 'The Arrogance of the Market: The Economy of the Kingdom between the Mediterranean and Europe', in G. Imbruglia (ed.) *Naples in the Eighteenth Century. The Birth and Death of a Nation State*. Cambridge: Cambridge University Press.

Sansa, R. 2000. 'Il mercato e la legge: la legislazione forestale Italiana nei secoli XVIII e XIX', in P. Bevilacqua and G. Corona (eds.) *Ambiente e risorse nel Mezzogiorno contemporaneo*. Corigliano Calabro: Meridiana Libri.

Santiago, Myrna. 2006. *The Ecology of Oil: Environment, Labor, and the Mexican Revolution, 1900-1938*. Cambridge and New York: Cambridge University Press.

Schabas, Margaret. 2005. *The Natural Origins of Economics*. Chicago: University of Chicago Press.

Scott, James. 1976. *The Moral Economy of the Peasants. Rebellion and Subsistence in Southeast Asia*. New Haven: Yale University Press.

Scott, James. 1998. *Seeing Like a State. How Certain Schemes to Improve the Human Condition Have Failed*. New Haven: Yale University Press.

Sereni, Emilio. 1948. *Il Mezzogiorno all'opposizione: dal taccuino di un ministro in congedo*. Torino: Einaudi.

Sereni, Emilio. 1997 [1966]. *History of the Italian Agricultural Landscape*. Cambridge and New York: Cambridge University Press.

Shepard, Paul. 1967. *Man in the Landscape. A Historic View of the Esthetics of Nature*. New York: A. Knopf.

Sieferle, Rolf P. 2010 [1982]. *The Subterranean Forest. Energy Crisis and the Industrial Revolution*. Cambridge: The White Horse Press.

Simmons, Ian G. 2008. *Global Environmental History*. Chicago: University of Chicago Press.

Smil, Vaclav. 1994. *Energy in World History*. Boulder: Westview Press.

Bibliography

Smil, Vaclav. 2008. *Energy in Nature and Society: General Energetics of Complex Systems.* Cambridge, MA: The MIT Press.

Spagnoletti, Angelantonio. 1997. *Storia del regno delle Due Sicilie.* Bologna: Il Mulino.

Stearns, Peter. 1993. *The Industrial Revolution in World History.* Boulder, CO: Westview Press.

Steinberg, T.L. 1986. 'An Ecological Perspective on the Origins of Industrialization'. *Environmental Review*, 4, 261–76.

Steinberg, Theodore L. 2000. *Acts of God. The Un-natural History of Natural Disaster in America.* New York: Oxford University Press.

Steinberg, T.L. 2002. 'Down to Earth: Nature, Agency and Power in History'. *American Historical Review*, 3, 798–820.

Steinberg, Theodore L. 2004 [1991]. *Nature Incorporated. Industrialization and the Waters of New England.* Cambridge and New York: Cambridge University Press.

Striano, Enzo. 1999. *Il resto di niente. Storia di Eleonora de Fonseca Pimentel e della rivoluzione Napoletana del 1799.* Napoli: Avagliano.

Tarrow, Sidney. 1967. *Peasant Communism in Southern Italy.* New Haven: Yale University Press.

Taylor, A. 1996. 'Unnatural Inequalities: Social and Environmental Histories'. *Environmental History*, 4, 6–19.

Taylor, Peter. 2005. *Unruly Complexity. Ecology, Interpretation, Engagement.* Chicago and London: The University of Chicago Press.

Temin, P. 1966. 'Steam and Waterpower in the Early Nineteenth Century'. *The Journal of Economic History*, 2, 187–205.

Tino, P. 1989. 'La montagna meridionale. Boschi, uomini, economie tra Ottocento e Novecento', in P. Bevilacqua (ed). *Storia dell'agricoltura italiana in età contemporanea*, vol. 1. Venezia: Marsilio.

Tino, P. 2007. 'Territorio, popolazione, risorse. Sui caratteri originali della storia ambientale italiana', *I Frutti di Demetra*, 13.

Vecchio, Bruno. 1974. *Il bosco e gli scrittori Italiani del settecento e dell'età Napoleonica.* Torino: Einaudi.

Venturi, Franco. 1969. *Settecento riformatore.* Torino: Einaudi.

Viazzo, Pier Paolo. 1989. *Upland Communities: Environment, Population, and Social Structure in the Alps since the Sixteenth Century.* Cambridge and New York: Cambridge University Press.

Villani, P. 1955. *Giuseppe Zurlo e la crisi dell'antico regime nel regno di Napoli.* Roma: Annuario dell'Istituto Storico Italiano per l'Età Moderna e Contemporanea, 7.

Villani, Pasquale. 1969. *Feudalità, riforme e capitalismo agrario.* Bari: Laterza.

Villani, Pasquale. 1973. *Mezzogiorno tra riforme e rivoluzione.* Bari: Laterza.

Villani, P. 1989. 'Lotte contadine, riforma agraria e questione meridionale', in P. Villani. *Società rurale e ceti dirigenti (XVIII–XX secolo).* Napoli: Morano.

Viscogliosi, A. 1988. 'I Boncompagni e l'industria', in *Trasformazioni industriali nella media valle del Liri in età moderna e contemporanea.* Isola del Liri: Rotary Club.

Von Salis Marschlins, Ulysses. 1795. *Travels Through Various Provinces of the Kingdom of Naples in 1789.* London: T. Cadell.

156

Von Tunzelmann, Nick. 1978. *Steam Power and British Industrialization to 1860*. Oxford and New York: Clarendon Press.

White, Richard. 1995. *The Organic Machine. The Remaking of the Columbia River*. New York: Hill and Wang.

White, R. 2004. 'From Wilderness to Hybrid Landscapes. The Cultural Turn in Environmental History'. *The Historian*, 66, 557–64.

Wilkinson, R., 1988, 'The British Industrial Revolution', in D. Worster (ed.) *The Ends of the Earth: Perspectives on Modern Environmental History*. Cambridge and New York: Cambridge University Press.

Williams, Raymond. 1975. *The Country and the City*. New York: Oxford University Press.

Williams, Raymond. 1980. 'Ideas of Nature', in Raymond Williams. *Problems in Materialism and Culture*. London: Verso.

Withed, Tamara. 2006. *Forests and Peasant Politics in Modern France*. New Haven: Yale University Press.

Wohl, Ellen. 2004. *Disconnected Rivers: Linking Rivers to Landscapes*. New Haven: Yale University Press.

Worster, Donald. 1988. 'Studying Environmental history' in D. Worster, (ed.) *The Ends of the Earth: Perspectives on Modern Environmental History*. Cambridge and New York: Cambridge University Press.

Worster, D. 1993a. 'The Ecology of Order and Chaos', in D. Worster. *The Wealth of Nature, Environmental History and the Ecological Imagination*. New York and London: Oxford University Press.

Worster, D. 1993b. 'Restoring a Natural Order' in D. Worster. *The Wealth of Nature, Environmental History and the Ecological Imagination*. New York and London: Oxford University Press.

Worster, D. 2006. 'Why we need environmental history', in M. Armiero (ed.) *Views from the South. Environmental Stories from the Mediterranean World*. Napoli: CNR.

Wrigley, E.A. 2006. 'The Transition to an Advanced Organic Economy: Half a Millennium of English Agriculture'. *Economic History Review*, 3, 435–80.

Zarlenga, R. 1854. 'Brevi considerazioni sul progetto di rendere navigabile il fiume Liri'. *Poliorama pittoresco*, XV.

Zonderman, David A. 1992. *Aspirations and Anxieties. New England Workers and the Mechanized Factory System, 1815–1850*. New York and Oxford: Oxford University Press.

Notes

1. On the Industrial Revolution as a major ecological transition see, for example: Debeir et Al 1991 [1986], Clapp 1994, Crosby 2006, Fischer-Kowalski and Haberl 2007, Foster 1999, Krausman et Al 2008, Martinez Alier and Schandl 2002, McNeill 2000, Sieferle 2001 [1982], Smil 1994, Simmons 2008, Steinberg 1986, Wilkinson 1988.

2. On the post-modern critique of economic history concepts see for example Mitchell 2000.

3. This definition of political economy is from Donald Worster. See Harvey (M.) 2008. On nature and political economy see also Harvey (D.) 1996.

4. On nature's agency in history see Steinberg 2002. On the concept of place see for example Casey 1997; for its use in environmental history see Flores 1996; see also Worster 2006. On local–global connections see Hannerz 1996.

5. See Bevilacqua 1989.

6. On deforestation and river degradation in the nineteenth century Mediterranean see McNeill 1992 and Radkau 2008. On Italy see Armiero 1999 and 2006, Hall 2005, Bevilacqua 1996.

7. See, for example, Bevilacqua and Rossi Doria 1984, and Bevilacqua 2010; see also Tino 1989, Palmieri 2000 and 2010, Sansa 2000, Gaspari 2000, Vecchio 1974.

8. The concept of Italy's 'original characters', referring to the evolution of the natural environment in the *longue durée*, first appeared as the title of Volume 1 of the Einaudi series *Storia d'Italia* in 1972 (see in particular Gambi 1972). For a more recent discussion of the role of nature in Italian history see the debate in *I Frutti di Demetra* (Bevilacqua 2005, Tino 2007, Cazzola 2006). On Italy's environmental history see also Armiero and Hall 2004.

9. See in particular Bevilacqua and Rossi Doria 1984, pp. 39–48. See also Bevilacqua 2000.

10. A perspective environmental historians owe to Donald Worster's landmark study, *Rivers of Empire* (1985). On industrial rivers in Europe and North America see, for example, Steinberg 1991, Brueggemeier 1994, White 1995, Platt 2002, Paavola 2002, Cioc 2002.

11. See McNeill 2000, pp. 189–239.

12. On risk and disaster in environmental history see, for example, Steinberg 2000; on floods in particular see Orsi 2004, Platt 2002, Kelman 2003, Saikku 2005. On the concept of social costs see Kapp 1971 [1950].

13. On capitalism's 'ecological contradictions' see O'Connor 1998, esp. pp 158–77; on eco-Marxism see also Benton 1996.

14. On social inequalities as a key issue in environmental history see Taylor 1996 and Steinberg 2002. On the social perception of nature see Williams 1980; on the co-constitution of economy and culture see also Cosgrove 1998, pp. 54–68. On culture, nature and historical materialism see O'Connor 1998, pp. 29–47. With different approaches, most environmental historians have investigated changing ideas of nature

as these relate to material environmental change. For some of the theoretical positions on such matters see Worster 1988, Merchant 1989, pp. 1–26, and Merchant 2004, Cronon 1992, White 2004. Political Ecology is also much concerned with the analysis of environmental discourses and their relationship to environmental practices: see for example Peets and Watts, 1996.

15. See Von Salis 1795, pp. 294–7.

16. See De Seta 1982; Black 2003.

17. S. Mazzella, *Descrittione del Regno di Napoli*, Napoli: Cappelli 1601, p. 23; quoted in Carbone 1971, p. 406.

18. G.B. Pacichelli, *Il Regno di Napoli in prospettiva diviso in dodici Province*, Napoli: Mutio 1703, pp. 121 and 144; quoted in Carbone 1971, p. 407.

19. N. Amenta, *Capitoli*, Firenze 1721; ibid.

20. L. Giustiniani, *Dizionario Geografico Ragionato del Regno di Napoli*, Napoli: Manfredi 1797, pp. 177–183; quoted in Carbone 1971, p. 408.

21. See Gutwirth 1978.

22. See the city booklet *Città di Isola del Liri*, Formia: M & P Edizioni 2002.

23. The term is used by Denis Cosgrove to describe Claude Lorrain's depiction of woods in 'Landscape with Ascanius shooting the stag of Silvia'; see Cosgrove 1998, p. 158.

24. The clearest influence on Bidauld's style is that of Henry Valenciennes – probably the most important landscape artist of the time – who had spent several years in Italy shortly before Bidauld: his paintings, mostly of the Roman Campagna, expressed a feeling for nature that was 'silent, enchanted, antisublime'. See Ottani Cavina 2004, p. 192.

25. Founders 1975, p. 317.

26. See also http://cartelfr.louvre.fr/cartelfr/visite?srv=car_not_frame&idNotice=18908

27. I borrow this expression from the title of Arnold 2006.

28. Also considering that the Roman Campagna had been a preferred theme for Claude Lorrain, the master of Arcadian landscape painting, a century before. See Cosgrove 1998, pp. 157–60.

29. See Scott 1998, p. 3; see also ibid., 'Part 1. State Projects of Legibility and Simplification', pp. 9–85.

30. See, for example, the stories of the Irk (Platt 2005), the Ruhr (Brueggemeier 1992), and the Rhine (Cioc 2002, Blackbourn 2006).

31. See also Di Biasio 1997.

32. My translation from the Wikipedia entry for 'accumulazione'. See http://it.wikipedia.org/wiki/Accumulazione

33. In the remainder of the book, capitalised type will be used when referring to the discipline of Political Economy.

34. See Merchant 2004, p. 80. On the Improvement ideal see also Meiskins 1999, pp. 89–112 and Cosgrove 1998, p. 224.

35. See Polanyi 1944, p. 33.

36. For a beautiful narrative of the absolutist State's war on nature see Blackbourn 2006, pp. 21–75.

37. Historian Eric Hobsbawm coined the expression 'Age of Revolution' to indicate the period of European history marked by the effects of both the French and the Industrial Revolutions (roughly 1789–1848): see Hobsbawm 1996 [1962]. Despite undergoing the fate of many general historical labels under the straws of post-modern critique, this expression is maintained here as perfectly suited to the narrative and time-frame of this book.

38. Gaetano Filangieri's treatise on the Science of Law [*Scienza della Legislazione*], written in the aftermath of the American Revolution and also the result of extensive international correspondence, is considered the best example of how the Neapolitan philosophers were participating in the broader context of the Age of Revolution. See Venturi 1969.

39. See Salvemini 2000 and 1981.

40. See Salvemini 2000, pp. 46–47.

41. On Genovesi and the Neapolitan Enlightenment School see also Venturi 1969 and Robertson 1997.

42. Salvemini 2000; see also Davis 1979 and Macry 1974.

43. See Davis 1979.

44. See Davis 2006, p. 19.

45. Ibid.

46. Although he was widely read in the field, especially in Scottish and English philosophers, he was apparently acquainted neither with the physiocrats nor, on the natural science side, with Linnaeus. Coming twenty years before the publication of *The Wealth of Nations*, Genovesi's use of the term '*economia*' must be thus considered an original contribution to the development of the discipline of Political Economy.

47. One of Genovesi's first works was an introduction to the Neapolitan edition of P. von Musschenbroek's *Elementa Physicae* (1745): a long 'physical-historical dissertation on the origins and constitution of things', with ample references to the classics, where Genovesi expressed his strong belief in the need for philosophy to abandon abstract speculation in favour of the scientific method, thereby contributing to the advancement of all practical aspects of social life. See De Sanctis 1986.

48. The Cartesian method had been known in Naples since the mid seventeenth century, when a 'modern' strand of mathematics and the corpuscular conception of matter had been established, against the still powerful presence of the Inquisition and censorship. Newtonianism spread even more quickly in the capital city, through the work of Celestino Galiani and of Nicola De Martino, a university professor who – in the 1720s– introduced the *Principia* to students in Naples, including Genovesi.

49. De Sanctis 1986, pp. 21–27; see also Brancaccio 1988.

50. See Harvey 1989.

51. The Museum can be considered one of the *loci* where nature, politics, and economics spatially intertwined with each other in nineteenth century Italian history: it was installed in 1801 within a former Jesuit college, 24 years after the Bourbons had banned the order from the kingdom; in 1845, the Museum would host a gigantic

(1600 participants) congress of Italian scientists, which marked an important moment in the formation of a national Italian identity; in 1848, it was chosen as the headquarters of the Neapolitan republican parliament, thereby making its entry into the history of the Italian Risorgimento. See De Sanctis 1986, pp. 69–78.

52. See 'Delle lezioni di commercio, o sia d'economia civile', in Genovesi 1977, p. 174 (my translation).

53. The name 'Greater Greece,' indicates the areas of southern Italy that were colonised by Greek settlers starting from the eight century BC.

54. See 'Discorso sopra il vero fine delle lettere e delle scienze', in Genovesi 1977, pp. 59–60.

55. Ibid., p. 60.

56. See Assante and Demarco, 'Introduzione' in Galanti 1979 [1794], p. ix.

57. Ibid., p. xvi–xvii. Galanti was impressed by the work of the German philosopher A.F. Busching, whose first four volumes of *Erdebeschreibung,* concerning the physical geography of Europe (1754–61), are considered to form a masterpiece of his time. Galanti devoted himself to updating the parts concerning Italy, editing a revision of the Italian translation in 1781. On Busching see http://www.1911encyclopedia.org/ Anton_Friedrich_Busching and http://www.britannica.com/EBchecked/topic/86065/ Anton-Friedrich-Busching.

58. Assante and Demarco, 'Introduzione', p. xviii.

59. Ibid., p. 6.

60. So Galanti stated in the preface to the book, which started with the assumption that 'after so many marvellous discoveries have been made about the systems of the world, improving history, navigation and geography', all European nations had finally devoted themselves to 'cultivat[ing] public economy and to perfect[ing] the science of government'. This, in fact, seemed to be the aim of knowledge itself: 'improving our life'. Galanti, 1979 [1794], p. 3.

61. Galanti was accused of having been inspired by Baron d'Holbach's *Sistema naturae*; however, he claimed he had never read the book. Ibid, p. xxxix.

62. This is clear even in simply considering the space that the author devotes to examining general political and economic matters (the revenue and the judiciary system, above all) and to the history of the kingdom, compared with that devoted to the physical and material description of the country. The geography part proper is condensed into an introductory chapter of a few pages – immediately followed by a long historical essay.

63. Ibid., p. 8.

64. Ibid., p. 9.

65. Strabo, Diodorus Siculus, Seneca and Pliny. Ibid., p. 10, notes.

66. Ibid., p. 13.

67. Ibid., p.123.

68. Ibid., p. 154.

69. Ibid., p. 239.

70. See for example De Lorenzo 2001a.

71. See Furet 1989, p. 688.

72. See Villani 1973, p. 164–65.
73. Ibid., p. 165.
74. On the 1792 Edict see Corona 1995.
75. Davis 2005, p. 54.
76. See De Negri 1992.
77. Davis 2005, p. 76.
78. See Pistilli 1824, p. 17.
79. Davis 2005, pp. 78–81. See also Robertson 2000.
80. See De Lorenzo 2001b.
81. On Eleonora Pimentel Fonseca see Striano 1999 (translated into French and German); see also Macciocchi 1993.
82. In the early twentieth century, Cuoco's harsh self-criticism was echoed by both the philosopher and historian Benedetto Croce (also a Neapolitan) and by Antonio Gramsci, each reinterpreting the Neapolitan 'passive revolution' as explaining the flaws of the Italian political unification process (Risorgimento). For a review of the historiographical debate see Davis 2005.
83. See Pistilli 1824. See also Lauri 1914 and Rizzello and Campagna 1988.
84. Pistilli 1824, p. 22.
85. I have echoed here the title of Amita Baviskar's 'Written on the Body, Written on the Land: Violence and Environmental Struggles in Central India', in Peluso and Watts 1998.
86. See Peluso and Watts 1998, p. 5.
87. See Armiero and Palmieri 2002.
88. See Corona 1995
89. See Pistilli 1824, p. 23.
90. *Il Sottintendente del Distretto di Sora all'Intendente*, Sora 21 Nov. 1806: ACS, IAC, 2353.
91. *Benedetto Marra, Procuratore dell'infelice città di Sora, al Re*, Sora, 12 Nov. 1806, ibid.
92. *Corrispondenza tra il Sottintendente e l'Intendente*, Oct. 1813: ACS, IAC, 2356.
93. See Pistilli 1824, p. 5.
94. See Merchant 2004, pp. 75–84.
95. See Palumbo 1979, pp. 6–10.
96. On the social meanings of waterpower in medieval Europe see for example Debeir et al. 1991 [1986]; see also Bloch 1967.
97. To Foucault, the term signified three interrelated process, all relevant to the case of southern Italy: 1) 'the ensemble formed by the institutions, procedures, analyses, and reflections, the calculations and tactics' allowing the exercise of state power, namely that 'which has as its target population, as its principal form of knowledge political economy, and as its essential technical means apparatuses of security'; 2) the tendency allowing the long term pre-eminence in western society of this kind of power over others, producing specific forms of governmental apparatuses and of knowledge; 3) the historical process which transformed medieval jurisdictions into state governments. See M. Foucault, 'Governmentality', in Rabinow and Rose 2003, p. 244.

98. See Davis 1994, p. 693.

99. See D'Elia 1992, pp. xxiv–xxv.

100. Davis 1994.

101. On the new cadastre in *Terra di Lavoro* see for example Migliore 2006

102. See Hobsbawm 1990, p 9.

103. De Sanctis 1986, p. 25.

104. Ibid., p. 64.

105. See Schabas 2005, p. 44.

106. The monumental work was coordinated by the Neapolitan botanist and director of the Orto Botanico Michele Tenore (1780–1861); the last volume was published in 1838. See De Sanctis 1986, pp. 79–89.

107. See her 'Introduzione' in D'Elia 1992; see also D'Elia 1994.

108. See Di Biasio 2004.

109. *Il Ciamberlano di Servizio presso S.M., duca d'Andria, al Ministro dell'Interno*, Naples (Camera del Re), n.d. and *Il sindaco di S.Germano all'intendente*, S.Germano 17 Oct. 1806: ASC, IAC 2353.

110. See Spagnoletti 1997.

111. *Benedetto Marra, procuratore dell'infelice città di Sora, al re*, Sora, 12 Nov. 1806: ASC, IAC 2353.

112. Ibid.

113. *Il sottintendente del distretto di Sora all'intendente*, Sora 21 Nov. 1806, ibid.

114. See D'Elia 1992, p. xxi.

115. Ibid., p. vii.

116. See Desrosières 1998 [1993].

117. The districts were defined as areas large enough to be of some significance but also designed in a way that people could reach the main town in a single day from anywhere within the district. Ibid., p. 32.

118. Ibid., p.35.

119. Ibid., p. 41.

120. Ibid., p. 44. On Italy see Patriarca 1996.

121. See D'Elia 1992, p. xvi. On Cagnazzi see Salvemini 1982; see also Patriarca 1996, pp. 27–28.

122. See Bevilacqua 1996, pp 196–214.

123. See De Marco 1988, pp. 70–72.

124. See G. Nicolucci. 'Sopra un cranio preistorico rinvenuto presso Isola del Liri (Terra di Lavoro)', in Carbone 1971, pp. 122 ff.

125. See De Marco 1988, p. 84.

126. Ibid., p. 387 and 479.

127. Ibid., p. 391.

128. Ibid., p. 276.

129. Ibid., p. 254.

130. Ibid., p. 520.

131. Ibid., p. 525.

132. *Corrispondenza tra il Sottintendente e l'Iintendente*, Oct. 1813: ASC, IAC 2356.

133. *Il Sottintendente all'Intendente*, Sora 17 May 1813, ibid.

134. *Il Sottintendente all'Intendente*, Sora 7 May 1814, ibid.

135. *Il Sottintendente all'Intendente*, Sora 30 Aug. 1813, ibid.

136. *Il Sottintendente all'Intendente*, Sora Dec. 1811, ibid.

137. See Palmieri 2000. On scientific forestry in the nineteenth century see Scott 1998, pp. 11–12; see also Guha 2000, pp. 36–44.

138. For a critique of environmental declension narratives see Cronon 1992 and 1996; see also Merchant 2004 and Radkau 2008 [2002].

139. Quoted in D'Elia 1992, p. 3 and 313.

140. A number of ancient towns (Capua, Taranto, Benevento, Cuma, Brindisi, Sibari and Metaponto among others, all of pre-Roman foundation) were said to have already regained their old prestige, 'rising from their ruins, brighter than ever'. Ibid.

141. Ibid., p. 5.

142. Ibid., p. 71–72.

143. Ibid., p.72.

144. For an anthology of eighteenth to twentieth century texts see Bevilacqua and M. Rossi Doria 1984. On environmental degradation in nineteenth century Italy see also Bevilacqua 1996.

145. See Vecchio 1974, Corona 1995 and Corona 2004; see also Di Biasio 2004, pp. 103–104.

146. 'Monticelli, a Benedictine, was for two years (1792–1794) professor of ethics at the University of Naples. The following years till 1800 were spent in prison as a result of his participation in the political disturbances of the time. He was appointed professor of chemistry at the University of Naples in 1808. His large collection upwards of 2000 specimens of Vesuvian products both minerals and lavas etc was purchased in 1823'. See British Museum 1904. Reference to some of Monticelli's publications about the Vesuvius can be found in Royal Society of London 1870, p. 458.

147. See Monticelli, 1809, p.1.

148. Ibid., p. 7.

149. Ibid., p. 11.

150. Ibid., p. 23.

151. See V. Cuoco *Rimboschimenti e Bonifiche*, now in Bevilacqa and Rossi Doria 1984, p. 171.

152. See Villani 1955.

153. See Dias 1841.

154. *Circolare del Gran Giudice Ministro della Giustizia del 13 settembre 1809*: ASN, MI 2° Inv. 543 bis, 130.

155. See for example McNeill 1992.

156. See Radkau 2008 [2002], p. 134. See also Rackam 2006.

157. See Radkau, 2008 [2002], p. 132.

158. See Gallo 2002 and Henneberg 1998.

159. See Sereni, 1997 [1966] pp. 17–21.

160. Though reaffirming the devastating impact of barbarian invasions after the fifth century, Sereni's picture of environmental decline in the South was in turn much more concerned with the elements of what Antonio Gramsci had called the internal 'disgregation' of Roman agrarian society. Again, a declensionist narrative of southern Italy was reinvented in relation to a very different political project – that of the post-war Italian communist party: see also Sereni 1948 and Tarrow 1967.

161. First theorised by Carolyn Merchant in her landmark study on New England, Ecological Revolutions are major transformations of society–nature relationships geared on irreversible changes occurring in three interrelated spheres: production (how nature and work are mixed and to what end), re-production (biological and social), and consciousness (how people see, imagine and understand nature). See Merchant 1989, especially pp. 1–26.

162. See for example Stearns 1993.

163. The literature on proto-industry has a long story, including analysis from Marx, Sombart and Vidal de la Blache. It saw a period of renewed interest during the 1970s with the works of F. Mendels, C. and R. Tilly; I refer here to the landmark study of Kriedte et al. 1981 [1977], also including references to previous debates. See also the more recent Pollard 1997.

164. On the capitalist vision of waterpower in the New England case see Steinberg 1991, pp. 71–76.

165. See for example Viazzo 1989; see also Ramella 1983.

166. On the process of fulling woollens see Hunter 1979 [1985], pp. 21–22. On the connection between proto-industry and waterpower see Pollard 1997, p. 230 ff.

167. Early documents regarding the introduction of hammers and fulling machines in the Duchy of Sora – then a feudal domain of the Della Rovere family – date from the period between 1539 and 1556; from 1580 the Boncompagni improved the art of wool by calling a *Maestro* from Florence; a few years later Duke Giacomo purchased a paper mill in Carnello from another unspecified Florentine. Furthermore, in order to increase the manufacturing of woollen clothes, in 1621 the King of Naples decreed the free trading of raw wool with the territory of Isola Liri. In the mid eighteenth century, Duke Gaetano restored the ancient *valche* of Carnello and called for the immigration of Dutch artisans to implement the art of wool. See Corradini 2004; see also Rizzello and Campagna 1988 and Associazione Alunni del Tulliano, *Arpino: storia arte e cultura*, http://www.laciociaria.it/comuni/arpino_moderna.htm By the late feudal pereiod, however, the Dukes had to compete with the merchants to exercise their control over water: see Viscogliosi 1988.

168. *Il Decurione Luigi Guarnieri al Minstro dell'Interno*: s.d. (1807?): ASC, IB, PS 44, 81

169. On the role of waterpower in the technology of the first industrial revolution see: Temin 1966, Gordon 1983, Reynolds 1983, Hunter 1979, Hills 1970, Ciriacono 2006b.

170. *Sentenza della Commissione Feudale tra l'Università di Arpino e l'Amminstrazione di Casa Reale,* 17 Feb. 1810: ASN, BSCN, vol. 2, p. 611.

171. *Il Sottintendente del distretto di Sora al Ministro dell'Interno,* Naples 31 Dec. 1815: ASN, MI 2nd inv. 566; and *Il sottintendente Corcioni all'intendente,* Sora 18 December 1818: ASC, IB, AC 2359. On the woollen industry in southern Italy see also De Matteo 1984 and De Majo 1990.

172. See De Marco 1988, p. 525.

173. See Dell'Orefice 1988, p. 118; a more recent version of the 'endless' machine can still be seen in Isola within the abandoned 'Cartiera Boimond', now entrusted to the local branch of the organisation Italia Nostra.

174. See De Matteo 1984.

175. *Il Sottintendente all'Intendente,* Sora 23 July 1830: ASC, IB, AC 2359.

176. *Il Sottintendente all'Intendente,* Sora 13 November 1831: ASC, IB, AIC 4.

177. Ibid. See De Majo 1990, p. 85.

178. Ibid., p. 112.

179. Ibid., p. 90.

180. See Reynolds 1983, p. 7.

181. See White 1995.

182. See Hunter 1979, pp. 205–11.

183. See for example Gordon 1985. See also Hunter 1979, pp. 205-11.

184. See also Zondermann 1992, pp. 31 ff.

185. See Cimmino 1986, pp. 186–87.

186. *Francesco Lanni all'Intendente,* Sant'Elia 19 Feb. 1853, and *Il Ministro dell'Interno all'Intendente,* Naples 5 July 1854: ASC, IB, AIC 6–7.

187. See Finotti 1891; for a more detailed account on the elabouration of energy data see Barca 2004.

188. On these aspects see Mathias 1999; Benoit 1985; Debeir et Al. 1986.

189. ASC, IB, AP 121.

190. For a comprehensive survey see Stearns 2007, especially Chapter 4.

191. See De Majo 1990.

192. See Mantoux 1999 [1909], pp. 67–69 and 233 ff.

193. On these aspects see Greenberg 1982 and Greenberg 1990; see also Berg 1980.

194. See De Majo 1990 p. 121.

195. For a definition of 'agrarian environment' see Agrawal and Sivaramakrishnan 2001.

196. See Dewerpe 1986, p. 94

197. On the causal relationship between increase of agricultural productivity and industrialisation in the British and Western European case see the classic Jones 1974; for a more recent reassessment see Wrigley 2006.

198. See F. Cirelli. 1858. *Il Regno delle Due Sicilie descritto ed illustrato.* Naples, quoted in Carbone 1971, pp. 201 ff.

199. See Mancini 1992, pp. 229 ff. See also Protasi 2002, pp. 57–59.
200. On these aspects, see in particular Chapter 5.
201. On spatial segregation as related to industrial watercourses see Platt 2005, esp. p. 196 ff. On the contradiction between production and reproduction see Merchant 1989, pp. 14–17 and 240–42.
202. See Ferri 1986.
203. See Isernia 1986 and Cimmino 1986b.
204. See Cimmino 1986a p. 147-48.
205. Ibid., p. 151-52.
206. Ibid., pp. 72 ff.
207. See Protasi 2002, p. 288.
208. See Romanelli 1819, p. 118.
209. Ibid., pp. 122–23.
210. See Cuciniello and Bianchi 1971 [1829].
211. See Fazzini 1836, p. 91.
212. See Zarlenga 1856, pp. 10–11.
213. See Lauri 1914, 9–11.
214. For an interesting comparison on this aspect see Santiago 2006. See also Harris 2003.
215. See Lauri 1914, p. 11.
216. Ibid., p. 108.
217. See Marx 1964, p. 26.
218. For a comparison with the North American case see Nye 2005.
219. See Williams 1975.
220. On landscape and the tourist gaze see Shepard 1967, pp. 127 ff.
221. See Wrigley 2005; for an Italian perspective see Federico and Malanima 2004.
222. Classical examples of this view are Landes [1969], and North and Thomas 1973. About the southern Italian case see Davis 1979.
223. For such a perspective see Debeir et al. 1991 [1986].
224. For a narrative of inequality and environmental risk as related to industrialisation in the 'classic' Manchester case, see Platt 2005.
225. On peasant economy and resistance see the landmark Scott 1976; for an interesting comparison with another Italian case of peasant-workers see Holmes 1989.
226. Quoted in Cencelli 1920, p. 67.
227. For a history of the Two Sicilies see, for example, Spagnoletti 1997; in English, see, for example, Astarita 2005.
228. See, for example, Villani 1989; see also Andretta 2008.
229. See Davis 1994, pp. 695–96, and ff.
230. For an analysis of this issue from an environmental point of view see for example Armiero 1999 and Armiero 2008.
231. See Spagnoletti 1997, pp. 253 ff.; see also De Rosa 1979; Villani 1969.

232. See Spagnoletti 1997, pp. 221 ff. On brigandage see also Hobsbawm 2000 [1969].

233. On the discontent of southern Italy's 'progressive' elites with Bourbon rule after 1848 see Petrusewicz 1998, pp. 107–119. On the collapse of the Bourbon State in 1860 see Macry 2003.

234. Approximately 3500 brigands and 300 soldiers were killed between 1861 and 1863; see Bevilacqua 1997, pp 35–36.

235. 'L'Illustration, Journal Universel', numero Jan–June 1862. The interview has been published on the website http://www.brigantaggio.net/Brigantaggio/Briganti/Chiavone_intervista.htm translated by Armando Calvano.

236. As testified by a plaque erected in his honour in the village of Scifelli in May 2000. See http://www.brigantaggio.net/Brigantaggio/lapidi/chiavone/chiavone.htm#gentile

237. See Peluso and Armiero 2008, p. 12. See also Moe 2002, Dickie 1999.

238. See Petrusewicz 1998.

239. See Grossi 1997, pp. 191–99 and 275 ff.

240. Ibid., p. 213 ff.

241. The term is the Italian translation for the Latin word Meridies, indicating the South. Garibaldi popularised it during his 1860 campaign, and it soon became a synonym for socio-economic backwardness and 'otherness'. See Dickie 1999.

242. For a review of studies on this issue, see Lumley and Morris 1997.

243. See Petrusewicz 1996; see also Bevilacqua 1997, p. 16 and Spagnoletti 1997 pp. 254–55.

244. On this aspect in particolar see Bevilacqua 1989.

245. See Davis 1979.

246. See Bevilacqua 1997, pp. 15-20.

247. See for example Macry 2002 [1988]; see also Spagnoletti 1997, Bevilacqua 1997 and Petrusewicz 1998.

248. See Spagnoletti 1997, p. 229.

249. See Armiero 1999, especially Chapter 1; see also Spagnoletti 1997, p. 242.

250. See Palmieri 1993.

251. See for example Spagnoletti 1997, p. 260.

252. See De Rivera 1825.

253. On De Rivera see also D'Elia 1994, p. 30.

254. Ibid., p. 19.

255. Ibid., p. 74.

256. Ibid., p. 32.

257. Formed of three volumes published in the course of ten years, the book was printed on paper from the *Cartiera delle Forme* in Carnello. See De Rivera 1832–42.

258. Ibid., p. 19. For a broader analysis of the political economy of public works in the kingdom see D'Elia 1996.

259. See Petrusewicz 1998, pp.76 ff. and 85 ff.; see also Palmieri 1993.

260. On the Bourbon Bureau of Land Drainage and Impriovement see D'Elia 1994, pp. 103–06.

261. On the 'tragedy of the commons' debate see Hardin 1968, Ciriacy Wantrup and Bishop 1975, McCay and Acheson 1987. For an environmental history perspective see McEvoy 1986.

262. See Laghezza and Ranieri 1845, pp. 4–5.

263. See Romano 2005.

264. See Bianchini 1971 [1857], p. 495-96.

265. See De Majo 1990, p. 75.

266. ASN, MI 2° inv., 3102, 10.

267. See De Majo 1990, pp. 99 and 101.

268. *L'Intendente di Terra di Lavoro al Ministro dell'Interno:* 17 Jan. 1835, ASN, MI 2° inv., 3102, 10.

269. ASN, MI 2° inv., 543 bis, 73

270. *Delibera del Decurionato di Isola del Liri:* 4 Jan. 1839: ASC, IB, ADC 84.

271. Ibid.

272. *Il Ministro dell'Interno all'Intendente di Terra di Lavoro,* 18 Apr. 1835: ASN, MI 2° inv. 530, 18.

273. *Il Ministro dell'Interno all'Intendente di Terra di Lavoro,* 20 May 1835: ibid.

274. *Deliberazione del Decurionato di S.Elia,* 22 June 1835: ibid.

275. *Il Ministro dell'Interno all'Intendente di Terra di Lavoro,* 15 July 1835: ibid.

276. *Esposto dei Fratelli Picano al Consiglio d'Intendenza di Terra di Lavoro,* 25 July 1835: ibid.

277. *L'Intendente di Terra di Lavoro al Ministro dell'Interno,* 25 July 1835: ibid.

278. *L'Intendente di Terra di Lavoro al Ministro del'Interno,* 24 Mar. 1838: ASN, MI 2° inv. 3102, 1.

279. *Deliberazione del Decurionato di S.Elia,* 8 Apr. 1836: ibid.

280. *Rapporto dell'arch. Rendina al Consiglio d'Intendenza di Terra di Lavoro,* 26 July 1836: ibid.

281. ASN, MI 2° inv. 3102, 6.

282. *L'Intendente di Terra di Lavoro al Ministro dell'Interno,* Caserta 12 Sept. 1839: ASN, MI 2° inv. 26.

283. *Sentenza del giudice G.B. Ferrante nella causa tra Giovan Battista De Ciantis e la collegiata di S.Restituta,* Sora 30 Oct. 1839: ASC, IB, CA LXXI.

284. *Felice Viscoglisi al Ministro dell'Interno,* 7 Jan. 1857:ASN, MI 3° inv. 678, 120.

285. *L'Intendente di Terra di Lavoro al Ministro dell'Interno:* 9 Oct. 1857, ibid.

286. The villages were Conocchietta, Santa Rosalia, Sant'Andrea and San Domenico: *Vincenzo De Ciantis al prefetto di Caserta,* 10 Aug. 1866: ASC, PCA, inv. 2, 2180.

287. *Felice Viscogliosi al Prefetto di Caserta,* 7 Sept. 1871: ACS, PCA, inv. 1, 7247.

288. *Felice Viscogliosi al Ministro del Lavori Pubblici:* 3 July 1872, ibid.

289. *Progetto Di Napoli-Viscogliosi*, 6 Aug. 1870 and *Relazione dell'Ing. Capo dell'Ufficio Tecnico di Caserta sulle opposizioni del sig. F. Roessinger alla domanda del sig. F. Viscogliosi*, 16 Nov. 1872: ASC, PC, 1, 4.

290. *L'ingegnere capo del Genio Civile di Caserta al Prefetto*, 4 Mar. 1872: ASC, PCA, inv. 6, 5487.

291. ASC, PPS, 2161.

292. See Parisi and Pica 1996, p. 115. On the Fucino drainage see also Raimondo 2000.

293. The data refer to minimum flows, measured in the 1880s: see MAIC 1895, pp 31–32.

294. ASC, PCA, inv. 6, 5486.

295. ASC, PCA, inv. 6, 5482.

296. These were Mazzetti, Questa, Società delle Cartiere Meridionali, Giuseppe Sarra, Beniamino Viscogliosi, Angelo Simoncelli; later, other riparian landowners took part in the lawsuit. Ibid.

297. ASC, PPS, cat. XXII, 234, 2471.

298. *Corpo Reale del Genio Civile, Provincia di Caserta Circondario di Sora, Constatazioni di derivazione dal Fibreno e Liri, Verbale di visita locale redatto dagli ingg. Edoardo Mezzacapo e Francesco Silvestrini*: ASC, PPS, cat. XXII,. 233, 2467.

299. See CRGC 1907.

300. *Verbale di visita del 16 giugno* : ASC, PPS, cat. XXII, 233, 2467.

301. Ibid.

302. *Osservazioni che si presentano dai sottoscritti (sigg. Gemmiti) all'uffizio del Genio Civile di Caserta nel sopraluogo del 17 Giugno 1896*, ibid.

303. See in particular Ostrom 1993.

304. See Steinberg 1991.

305. ASN, MI, 2nd Inv., 530 and ASC, IB, AC, 15.

306. ASN, MI, CPTL 4054.

307. ASC IB, PS-B 44, 183.

308. ASC, PPS, 188, 2105.

309. ASC, PPS, cat. XXII, 253, 2547.

310. ASC, PPS, cat. XXII, 188, 2463.

311. ASC, PPS, cat. XXII, 253, 2547.

312. The other two main kinds were the 'classic' situation – that is depopulated swamps to be drained and returned to agriculture; and the 'intensive' water economies – those where bad drainage and semi-permanent flooding were caused by a mix of ill-regulated uses (irrigation, fishery, waterpower, hemp and flax maceration, rice cultivation). For this classification see D'Elia 1994.

313. See D'Elia 1994, p. 43.

314. Ibid., p. 74.

315. See S. De Renzi, *Miasmi Paludosi Contagi ed Epidemie*, Naples 1826, cited in D'Elia 1994, pp. 28-35.

316. *Il Direttore Generale di Ponti e Strade al Ministro dell'Interno,* 30 Nov. 1825: ASN, MI, 2° Inv., 530. The report added that the road bridge to the Abruzzi during floods acted like another dam, and that between Sora and its surrounding farmland was obstructed, while Sora was flooded.

317. *Il Direttore Generale di Ponti e Strade al Ministro dell'Interno,* 27 December 1825, ibid.

318. *Deliberazione del Decurionato di Sora,* 25 Feb. 1826: ibid.

319. *Supplica del Vescovo di Sora al Ministro dell'Interno,* 25 May 1826: ibid.

320. *Il Curato di Santa Restituta al Sindaco di Sora,* 24 May 1826: ibid.

321. Cfr. Il Vescovo di Sora, Aquino e Pontecorvo al Sindaco di Sora, 8 May 1826, ibid.

322. Ibid.

323. *La Madre Superiora del Monastero delle Clarisse al Sindaco di Sora,* 8 May 1826: ibid.

324. *Savino Marsella al Sindaco of Sora,* 26 May 1826: ibid.

325. *Il Direttore Generale di Ponti e Strade al Ministro dell'Interno,* 1 July 1826, ibid.

326. *Deliberazione del Decurionato di Sora,* 13 March 1827: ibid and ASC, IB, AC 2363.

327. *Il Sottintendente di Sora all'Intendente* di Terra di Lavoro, 21 May 1827: ibid.

328. *Il Ministro dell'Interno al Presidente della Consulta dei Reali Domini al di qua del Faro,* 12 Sept. 1827: ibid.

329. *Nota per lo Consiglio,* n.d. ibid.

330. See De Rivera 1832-42, vol. 1, pp. 112–14.

331. For an interesting comparison with France see Withed 2006.

332. See for example: Sereni 1997 [1967], Ciriacono 1998 and 2006a, Bigatti 1995 and 2000, Bevilacqua 2007 [1998].

333. On this matter see Barca 2006.

334. See Rombai, pp. 628–29.

335. For a thorough use of this concept in the case of German reclamations see Black-bourn 2006.

336. See for example Isenburg 2000, pp. 66–67.

337. Ibid., pp. 55–56.

338. Quoted in Palmieri 1993, p. xxi.

339. ASN, MI, 2° Inv. 530 and ASC, IB, AC 2356.

340. On the *ratizzo* system see D'Elia 1994.

341. See D'Elia 1996.

342. *Sessione del 1833*: ASN, MI, CPTL 4054.

343. *Sessione del 1842*: ibid.

344. *Il Sottintendente del Distretto di Sora all'Intendente Terra di Lavoro:* Sora, 30 Apr. 1852: ASC IB, PS-B 44, 183.

345. *Il Sottintendente del Distretto di Sora all'Intendente Terra di Lavoro:* Sora, 9 Jan. 1856: ibid.

346. *Il Ministro dei Lavori Pubblii all'Intendente di Terra di Lavoro:* Naples, 28 March 1857: ibid.

347. *L'Intendente di Terra di Lavoro al Ministro dei Lavori Pubblicñ,* Caserta 15 Apr. 1857: ibid.

348. *Il Ministro dei Lavori Pubblici all'Intendente di Terra di Lavoro,* Naples 5 Jan. 1858: ibid.

349. *Il Ministro dei Lavori Pubblici all'Amministrazione Generale delle Bonifiche,* Napoli 2 Dec. 1857: ibid.

350. *Il Sottintendente all'Iintendente di Terra di Lavoro,* Sora, 27 Oct. 1858: ibid.

351. *Il Sottintendente all'Intendente di Terra di Lavoro,* Sora, 29 Nov. 1858: ibid.

352. *Il Sottintendente all'Intendente di Terra di Lavoro,* Sora, 1 Dec. 1858: ibid.

353. *Il Ministro dei Lavori Pubblici all'Intendente di Terra di Lavoro,* Naples, 11 Dec. 1858: ibid.

354. *Il Direttore Generale delle Bonifiche all'Intendente di Terra di Lavoro,* Naples 5 Feb. 1859: ibid.

355. This had in fact already happened with the Torano works (4,000 ducats) and that for the lower catchment basin of the Volturno (10,000 ducats per annum, about 130,000 in 16 years, plus 546 ducats per annum 'to pay wages for the Land Reclamation Bureau employees'). *L'Intendente di Terra di Lavoro al Ministro dei Lavori Pubblici,* Caserta, 26 Feb. 1859: ibid.

356. Cfr. Sora, 30 Aug. 1861, *L'intendente del circondario di Sora al Governatore in Caserta,* in ASC, AP, b. 477, 5234.

357. See Ramazza 1996, p. 1.

358. *Il Direttore Generale dei Lavori Pubblici al Ministro,* Naples 18 Mar. 1863: ACS-R, LLPP, DGPAS 51b, 4.

359. ASC, PPS 188, 2105.

360. *La Deputazione Provinciale di Terra di Lavoro al Prefetto,* Caserta 20 July 1879: ibid.

361. *Il rappresentante del Principe di Torlonia, Gaetano Manetti, al Prefetto di L'Aquila,* Avezzano 4 Sept. 1879: ibid.

362. See Mancini 1882, p. 340.

363. *Il Sottintendente all'Intendente di Terra di Lavoro,* Sora, 22 January 1845: ibid.

364. *L'intendente di Terra di Lavoro al Sottintendente,* Caserta Sept. 1852: ibid.

365. *Andrea Tuzj al Ministro dei Lavori Pubblici,* Sora, 14 August 1870: ASC, PCA inv. 1, 7252.

366. *Andrea Tuzj al Ministro dei Lavori Pubblici,* Sora 20 January 1871: ibid.

367. *Il Sottoprefetto al Prefetto di Caserta,* Sora, 27 July 1871: ibid.

368. *L'ingegnere del Genio Civile Bianchi al Prefetto,* Caserta, 2 Mar. 1872: ibid.

369. *Il ministro dei Lavori Pubblici al prefetto di Caserta,* Rome 26 Oct. 1871: ibid.

370. ASC, PCA Inv. 3, 3109.

371. *La Deputazione Provinciale al Sindaco di Sora,* Caserta 28 June 1888: ASC, PCA Inv. 11, 10740.

372. *L'Ingegnere Capo del Genio Civile al Prefetto,* Caserta 8 Aug. 1900: ASC, PPS cat. XXII, 188, 2456.

373. *Il Sottoprefetto al Prefetto,* Sora 10 Nov. 1906: ASC, PPS cat. XXII, 188, 2463

374. ASC, PPS cat. XXII, 253, 2547.

375. Ibid.

376. See Dewerpe 1984.

377. Autorità di Bacino dei Fiumi Liri–Garigliano e Volturno. *Piano Stralcio per l'Assetto Idrogeologico, Rischio Idraulico: Bacino del Fiume Liri-Garigliano* (April 2006). See http://www2.autoritadibacino.it/

378. See Carbone 1971, pp. 365–66.

379. See Bellucci 2009, pp. 39–47.

380. For a thorough discussion of the dialectic between order and chaos in ecology see for example Taylor 2005. See also Merchant 1994, pp. 18–21, Prigogine 1994, Worster 1993a and 1993b.

381. See for example D'Souza 2006.

382. See Wohl 2004.

383. Comune di Isola Del Liri, Ufficio Relazioni con il Pubblico, *Il Parco Fluviale*: http://www.comune.isoladelliri.fr.it/newurp/urp/b.asp?etc=Ufficiandvar_b2=111

384. See for example Hall 2005.

385. Comune di Isola del Liri, *Il parco fluviale*.

Index

Index

Index

Index

Index

Lightning Source UK Ltd.
Milton Keynes UK
UKOW041422071112

201805UK00001B/152/P